INFRARED MICROSPECTROSCOPY

INFRARED MICROSPECTROSCOPY

Theory and Applications

edited by

Robert G. Messerschmidt

Spectra-Tech, Inc.
Stamford, Connecticut

Matthew A. Harthcock

Analytical and Engineering Sciences
Dow Chemical U.S.A.
Freeport, Texas

MARCEL DEKKER, INC. New York and Basel

Library of Congress Cataloging-in-Publication Data

Infrared Microspectroscopy : theory and applications / edited by
Robert G. Messerschmidt, Matthew A. Harthcock.
p. cm. -- (Practical spectroscopy : v. 6)
Includes index.
ISBN 0-8247-8003-5
1. Infrared spectroscopy. 2. Fourier transform spectroscopy.
I. Messerschmidt, Robert G. [date]. II. Harthcock, Matthew A.,
[date]. III. Series.
QD96.I5I56 1988
543'.0858--dc 19 88-10850

MARCEL DEKKER, INC.
270 Madison Avenue, New York, New York 10016

Current printing (last digit):
10 9 8 7 6 5 4 3 2 1

PRINTED IN THE UNITED STATES OF AMERICA

Preface

Infrared microspectroscopy, the coupling of optical microscopy and infrared spectroscopy, is not a new concept. In the 1940's instrumentation for recording infrared spectra on areas of a sample isolated using an optical microscope with all-reflecting optics was designed and limitedly utilized in a coupled mode with a dispersive infrared spectrophotometer. Until the early 1980's analysis of microscopic samples using infrared spectroscopy largely involved the use of what is referred to as "traditional" microsampling techniques. These "traditional" microsampling techniques involve, for example, pinhole masking or micro potassium bromide pellets.

In the 1970's a coupled Raman spectrometer and optical microscope became and continues to be a popular vibrational spectroscopic microprobe technique. Much has been written on the application of Raman microprobe techniques for obtaining vibrational spectra on microscopic particles as small as one micrometer. However, the continuous fluorescence problems associated with Raman spectroscopy have prevented broader application of the Raman microprobe.

In the early 1980's, a resurgence in the use of infrared microspectroscopy occurred for two basic reasons. First, the advantages (predominantly the multiplex and energy throughput advantages) of Fourier transform infrared (FT-IR) spectrophotometers versus dispersive instrumentation have allowed convenient and routine operation of an all-reflecting microscope with an infrared spectrophotometer. Currently, the diffraction limit of infrared radiation (10-20 micrometers) is the limiting condition for obtaining absolute spectra from a given area of a material. The second reason is trends in technology. Various fields of science are requiring different and specific answers to problems that "traditional" analytical instrumentation cannot provide, as will be demonstrated many times in this book.

The past several years have seen increasing attention to the field of infrared microspectroscopy. Various symposia have been held at meetings such as the Federation of Analytical Chemistry and Spectroscopy Societies (FACSS), the Eastern Analytical Symposium (EAS), the Pittsburgh Conference, and the Microbeam Analysis Society. The first book published on the topic appeared in August of 1987. The book, edited by Dr. Patricia B. Roush (Perkin Elmer Corporation), is based on a symposium chaired by Dr. Roush, containing nine papers on various areas of infrared microspectroscopy, and is titled *The Design, Sample Handling, and Applications of Infrared Microscopes* (ASTM Special Technical Publication 949, 1987).

This book of eighteen chapters is an outgrowth of symposia organized and held at FACSS XIII and EAS by the editors (FACSS-Harthcock, EAS-Messerschmidt) during the fall of 1986 and contains the theory behind the development of a high performance infrared microscope. Also included are the considerations necessary for using infrared microspectroscopy, for example, for imaging a sample and for coupling it with high performance liquid chromatography (HPLC). Finally, a variety of applications of the use of infrared microspectroscopy are contained in this volume.

The chapters in this book often address both specifics about a technique (e.g., sampling handling and new methods of using infrared microspectroscopy) and applications of the technique to various scientific fields (e.g., polymer characterization and semiconductor technology). The editors have chosen to classify the chapters in this book into the following categories:

- Instrumentation Considerations and Technique Advances in Infrared Microspectroscopy
- Analysis of Polymers by Infrared Microspectroscopy
- Applications of Polarized Infrared Microspectroscopy
- Application of Infrared Microspectroscopy to the Semiconductor Industry
- Application of Infrared Microspectroscopy to Biological and Pharmaceutical Research
- Miscellaneous Applications of Infrared Microspectroscopy

It will be apparent that several of the chapters overlap several of the general catagories above.

From the categories, one can see the impact the field of infrared microspectroscopy is having. The opening chapter by Messerschmidt explains the theory of operation and design of an FT-IR microscope. Topics such as Redundant Aperturing™ versus single aperturing, the numerical aperture, signal-to-noise ratio, coherence, etc. are discussed as they relate to the performance of an FT-IR microscope. Harthcock and Atkin reveal methodology using infrared microspectroscopy for obtaining images of a material based on functional group maps. Presented as a new imaging technique, examples of the application of this technique to polymer characterization are included.

Lang et al. in their chapter present considerations in keeping a clean working environment for doing infrared microspectroscopy. This chapter is more for the benefit of the new infrared or vibrational microspectroscopist who should be aware of contamination potential than for the optical microscopist who has been aware of contamination potential for years.

Polymer characterization is obviously an area that infrared microspectroscopy is influencing extensively. Humecki discusses sample preparation and analysis of polymers and contaminants in or from various matrices. Gerson and Chess describe the use of reflectance infrared microspectroscopy for delineating the failure of a polymer composite. Mirabella describes the simultaneous measurement of infrared and differential scanning calorimetry data to study changes in polymer structure which occur at various temperatures. The future of this technique for polymer characterization is also addressed by Mirabella.

The use of polarized infrared microspectroscopy is also addressed, in large part as it applies to polymeric materials. Chase discusses the details in obtaining accurate dichroic spectra from single polymer fibers. The advantage of using Redundant Aperturing in the measurement of dichroic spectra of single fibers is also described. Brasch and Lustiger describe the use of polarized infrared microspectroscopy to investigate the amount of tie molecules present in polyethylene. Hill and Krishnan provide examples of dichroic measurements on single crystals and polymers using infrared microspectroscopy.

The use of infrared microspectroscopy has been prescribed in the semiconductor industry for about the last seven years. Madden et al. describe how the technique is used to address contamination problems in the semiconductor industry. Krishnan shows the use of infrared spectra for determining the epitaxial layer thickness of a semiconductor and for determining contaminants in the wafers (e.g., oxygen, carbon).

In the chapter by Fuller and Rosenthal the potential of infrared microspectroscopy for studying biological specimens is shown. For example, the authors show how spectra can be obtained on a few red blood cells. The full potential of infrared microspectroscopy in the biological and medical sciences has definitely not been realized to date and it will certainly expand in these fields. Reffner shows in his chapter how polymorphic behavior, of pharmaceutical importance, can be investigated using infrared microspectroscopy. Reffner also includes an introduction which defines several terms used throughout this book.

Fraser et al. demonstrate the coupling of infrared microspectroscopy for problem solving in the industrial environment. They also identify the complementary nature of Raman microprobe spectroscopy and infrared microspectroscopy.

Schiering also provides various examples of the technique's use in an industrial or quality control environment. An example of the use of a diamond cell for flattening materials for analysis by infrared microspectroscopy is included and a nice brief history of infrared microspectroscopy is given. Sommer et al. describe the use of infrared microspectroscopy for studying paper chemistry.

The final chapter, by Wooton and Hughes, discusses the application of grazing angle reflectance infrared microspectroscopy to characterize the surface of metals which had been treated with various lubrication oils.

This book is intended to provide an overview of the theory and applications of infrared microspectroscopy. As observed from the contents and noted repeatedly by the authors, the influence of this technology has spanned many fields of science. The impact of infrared microspectroscopy is only starting to be realized. The material presented within the covers of this book provides a sampling of the data which can be obtained from this analytical technique.

The editors wish to thank each of the contributing authors for their chapters and their prompt response to issues related to getting this book to press. Also, the editors wish to thank the governing boards of the Federation of Analytical Chemistry and Spectroscopy Societies and the Eastern Analytical Symposium for their permission to publish this book, which is based on papers first given at their meetings. One of the editors (MAH) would like to express his appreciation to The Dow Chemical Company for allowing him the opportunity to serve as a co-editor of this book, his family for their understanding during the time it took to put this book together, and particularly his wife, Patricia, for assisting in some of the secretarial duties necessary to complete this book.

The efforts of those who have contributed in an editorial capacity should not go unmentioned. Most importantly, the editors wish to thank Ms. Jennifer Sting and Ms. Andrea Mikolowsky, who have completely produced this book (except for some of the figures) using Apple Macintosh® computers. Chapters received from the authors either were already in a machine-readable format or were brought into the Macintosh via an optical character reader. The chapters were then formatted into a standard Marcel Dekker page format using Aldus Pagemaker® software. Appropriate spaces were left for later insertion of figures, except in cases where the figure was Macintosh generated as is the case in the chapter by Messerschmidt. Equations were entered using A. Bonadio

Associates' Expressionist™ software. Indexing was done using Microsoft Excel® software. Camera-ready copy was output using an Apple Laserwriter® printer. Ms. Sting and Ms. Mikolowsky are to be congratulated for their hard work and perseverance through the production of a book such as this using state-of-the-art production techniques. Additional editorial assistance and proofreading were contributed by Mr. Donald Sting, Dr. John Reffner, Mr. Seth Lefferts, Mr. Robert Sebes, Mr. Thomas Mattone, and Ms. Joan Kwiatkoski. Mr. Lefferts also served as computer problem-solver. We extend our thanks to all mentioned.

The contributions of Spectra-Tech, Inc., to the publication of this work is also acknowledged. Spectra-Tech has donated the time of one of the editors (RGM), the services of both of the production editors (Ms. Mikolowsky and Ms. Sting), the time of many of the copy editors, as well as all computer time.

We would also like to thank the staff of Marcel Dekker, Inc., for their cooperation, most notably Ms. Rosemarie Krist and Ms. Vickie Kearn.

The editors would like to note that all chapters were received from the authors by no later than May 1987. All papers were accepted for publication no later than November 1987.

Matthew A. Harthcock

Robert G. Messerschmidt

Contents

Applications of Polarized Infrared Microspectroscopy

Application of Infrared Microspectroscopy to the Semiconductor Industry

Application of Infrared Microspectroscopy to Biological and Pharmaceutical Research

Miscellaneous Applications of Infrared Microspectroscopy

Contributors

Scott C. Atkin, Texas Applied Sciences and Technology Laboratories, Analytical and Engineering Sciences, Dow Chemical U.S.A., Freeport, Texas

Barbara Bergin, IBM Corporation, East Fishkill, Hopewell Junction, New York

Anthony S. Bonanno, Clarkson University, Department of Chemistry, Potsdam, New York

J. W. Brasch, Battelle Columbus Division, Columbus, Ohio

Bruce Chase, Central Research Department, E. I. DuPont de Nemours, Wilmington, Delaware

Catherine A. Chess, National Service Division, IBM Corporation, Danbury, Connecticut

David J. J. Fraser, Department of Chemistry, University of California, Riverside, California

Michael P. Fuller, Nicolet Analytical Instruments, Madison, Wisconsin

Dennis J. Gerson, National Engineering / Scientific Computing Center, IBM Corporation, Dallas, Texas

Peter R. Griffiths, Department of Chemistry, University of California, Riverside, California

Matthew A. Harthcock, Texas Applied Science and Technology Laboratories, Analytical and Engineering Sciences, Dow Chemical U.S.A., Freeport, Texas

Doyle M. Hembree, Jr., Martin Marietta Energy Systems, Inc., Oak Ridge, Tennessee

S. L. Hill, Bio-Rad, Digilab Division, Cambridge, Massachusetts

Randall L. Howell, Martin Marietta Energy Systems, Inc., Oak Ridge, Tennessee

Dennis W. Hughes, Ethyl Petroleum Additive Division, Ethyl Corporation, St. Louis, Missouri

Howard J. Humecki, Walter C. McCrone Associates, Inc., Chicago, Illinois

J. E. Katon, Molecular Microspectroscopy Laboratory, Department of Chemistry, Miami University, Oxford, Ohio

Nancy Klymko, IBM Corporation, East Fishkill, Hopewell Junction, New York

K. Krishnan, Digilab Division, Bio-Rad, Cambridge, Massachusetts

Patricia L. Lang, Molecular Microspectroscopy Laboratory, Department of Chemistry, Miami University, Oxford, Ohio

A. Lustiger, Bell Laboratories, Murray Hill, New Jersey

Karen Madden, IBM Corporation, East Fishkill, Hopewell Junction, New York

Robert G. Messerschmidt, Spectra-Tech, Inc., Stamford, Connecticut

Brian S. Miller, Molecular Microspectroscopy Laboratory, Department of Chemistry, Miami University, Oxford, Ohio

Francis M. Mirabella, USI Chemical Company, Rolling Meadows, Illinois

Kelly L. Norton, Department of Chemistry, University of California, Riverside, California

Joseph C. Oswald, Martin Marietta Energy Systems, Inc., Oak Ridge, Tennessee

G. E. Pacey, Molecular Microspectroscopy Laboratory, Department of Chemistry, Miami University, Oxford, Ohio

John N. Ramsey, IBM Corporation, East Fishkill, Hopewell Junction, New York

John A. Reffner, Spectra-Tech, Inc., Stamford, Connecticut

Robert J. Rosenthal, Nicolet Analytical Instruments, Madison, Wisconsin

David W. Schiering, Perkin-Elmer Corporation, Norwalk, Connecticut

Norman Smyrl, Martin Marietta Energy Systems, Inc., Oak Ridge, Tennessee

A. J. Sommer, Molecular Microspectroscopy Laboratory, Department of Chemistry, Miami University, Oxford, Ohio

David L. Wooton, Ethyl Petroleum Additive Division, Ethyl Corporation, St. Louis, Missouri

Instrumentation Considerations and Technique Advances in Infrared Microspectroscopy

1

Minimizing Optical Nonlinearities in Infrared Microspectrometry

ROBERT G. MESSERSCHMIDT *Spectra-Tech, Inc., Stamford, Connecticut*

1.1 INTRODUCTION

There are several fundamental concepts that should be understood in order to successfully perform FT-IR microspectroscopy which are unimportant in macro-spectroscopy. Many of these concepts are a result of optical principles which, to other than an optical physicist, may seem to contradict one's understanding of optics. In fact, some of the concepts necessary to describe the behavior of imaging in the infrared microscope *do* contradict classical geometric optics!

Most readers will be familiar with the concepts of the laws of reflection and refraction, and the concept of magnification. These are the basis of geometric optics, and can be summarized by one simple equation, Snell's Law of Refraction. This formula describes the bending of light as it passes through materials of differing refractive indices and past surfaces of various curvatures. A special case of this law for mirror surfaces states that upon the encounter of a "ray" of light with a reflective surface, the angle of incidence of that light ray with respect to the surface will equal the angle of reflection.

Many readers will also know about the common optical aberrations which can be predicted and proved through the application of geometric optical ray tracing. These, such as spherical aberration, chromatic aberration and coma (offense against the sine condition), are controllable. It is the optical designer's responsibility to keep these aberrations small enough so that they are unobjectionable in a given application.

This is as far as one routinely needs to go in the design of instruments to be used in the visible region of the spectrum, at low magification. However, in order to deal with the behavior of light when the eye is aided, for instance in a microscope system, then the diffraction of light energy must be considered. Diffraction occurs everywhere, not only in microscopes. Insofar as the eye is an optical instrument, the image of everything one looks at is altered by diffraction. Fortunately, given the acceptance angle of light into the eye and the wavelengths involved, the diffraction effect is below the resolution limit. Therefore, one does not notice the effect.

The diffraction effect in infrared microspectroscopy manifests itself as a blurring of the image information which one obtains in order to measure the spectrum of a given sample. Primarily, the spatial resolution of the measurement is affected. That is, the energy reaching the detector contains spectral information from a larger physical area than is expected or desired. This has it's obvious consequences. Spurious energy and/or spectral peaks can be present in the resultant spectrum, inviting quantitative and qualitative misinterpretation. Unfortunately, diffraction is a physical phenomenon, and it is therefore not possible to eliminate the problem. But there are design considerations that can minimize the effect.

1.2 GENERAL CONSIDERATIONS

The general layout of an infrared microscope is shown in Figure 1. It consists of transfer optics to bring the infrared radiation from the interferometer, imaging both the source image and the pupil image (the beamsplitter image) through the microscope. The visible optical train is parfocal and colinear with the infrared radiation. This is achieved in setting up the microscope by viewing through a series of small pinholes, and then aligning for infrared energy through the same pinholes. In actual use, the area of interest is brought to the center of the field of the microscope under visible (transmitted or reflected) light. Then the area to be analyzed is delineated with high contrast apertures in the remote image planes of the sample. Preferably these apertures are variable in size so that the area of interest may be "zeroed in on". Since the optical geometry is the same for the visible evaluation and the infrared detection, the diffraction

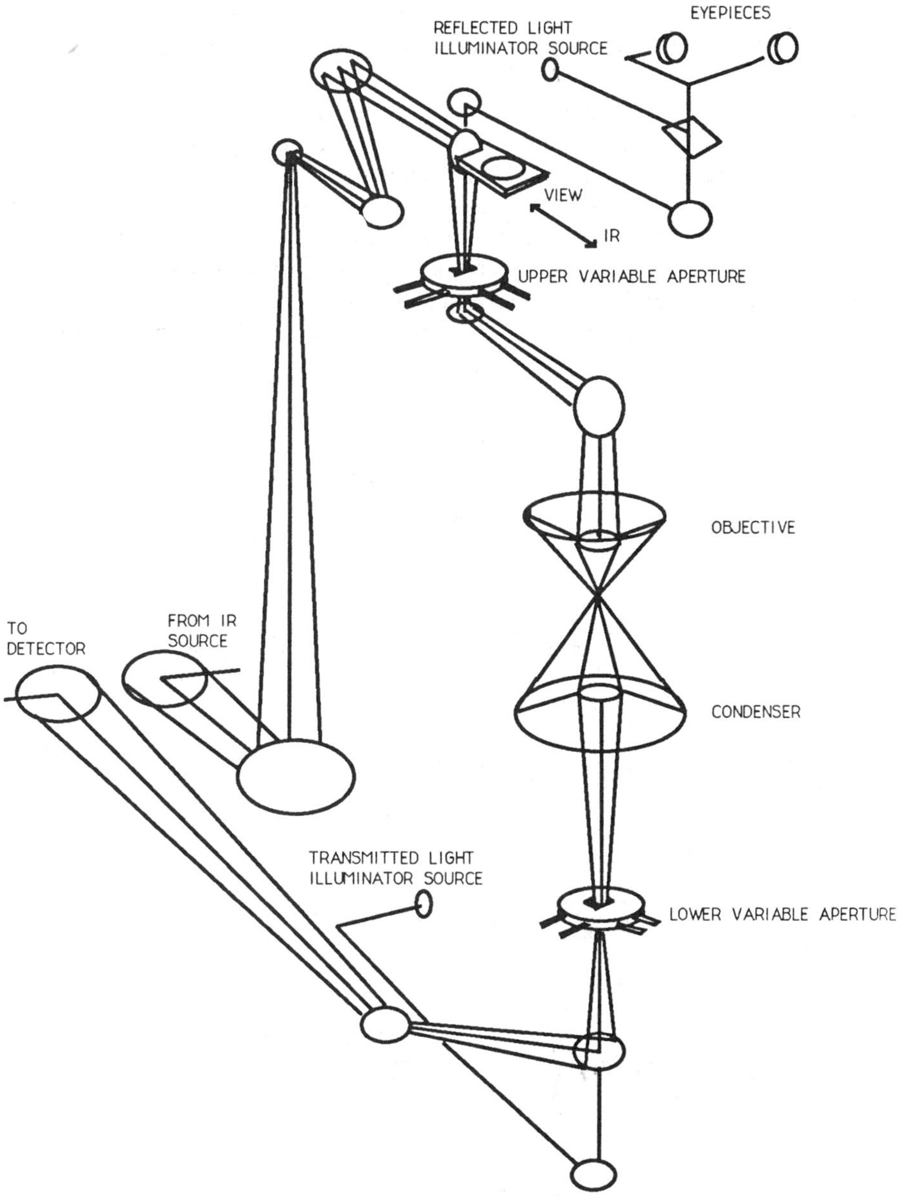

Figure 1. The general layout of an FT-IR microscope.

effect is worse for the detection step, because the wavelength is longer. It follows that one can see the area one wants to measure more clearly than can actually be measured. For FT-IR microscopy therefore, the spatial resolution is defined as the ability to measure the spectrum from an object delineated by the apertures without impurity radiation from neighboring objects. Since the infrared spectrum is quite complicated, consisting of thousands of data points, it is often difficult to say whether spurious energy has affected the result. If there is impurity, it can take the form of either spurious *narrow band impurity* from a nearby absorber, or spurious *broad band impurity* from a nearby hole. Obviously this plays havoc with the quantitative accuracy of the measurement. Both quantitative accuracy and signal-to-noise may be affected by the ability of a microscope to spatially resolve the specimen.

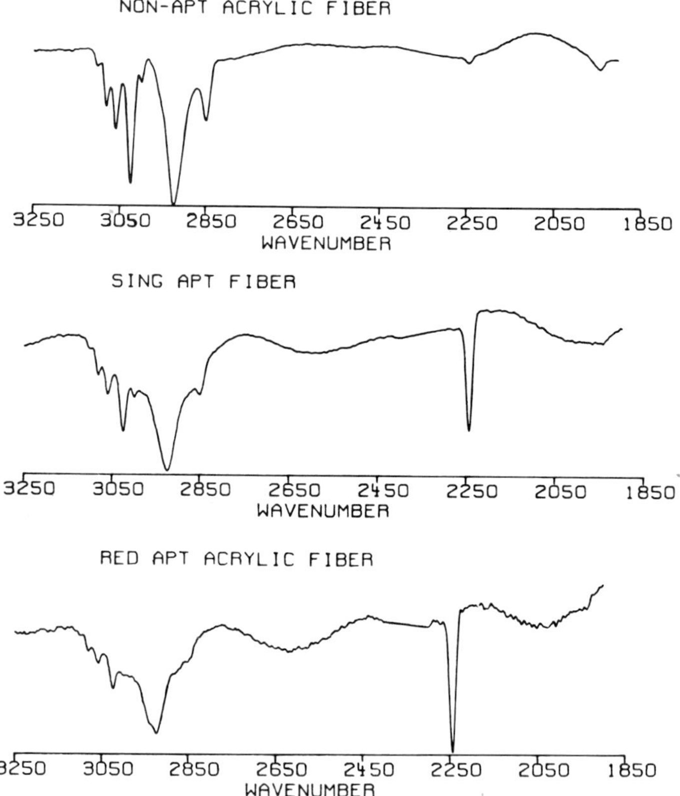

Figure 2. A demonstration of the problem of stray or spurious energy in an FT-IR Microscope. The specimen in all three spectra is an acrylic fiber embedded in polystyrene. Top, wide apertures. Middle, single aperture defining the fiber area. Bottom, redundant apertures defining the fiber area.

An interesting example is shown in Figure 2. An acrylic fiber, 20 micrometers in diameter, is embedded in polystyrene and then cross-sectioned to a 10 micrometer thickness. Three spectra are recorded, all with the same spectrometer conditions and for the same length of time. In the first spectrum, no particular effort is made to aperture out the surrounding embedding medium. In the second, one variable aperture is used to delineate the fiber. In the third spectrum, two apertures, one in the rear focal plane of the objective, and one in the rear focal plane of the condenser both delineate the same area of the sample. Quite obviously the first spectrum has the best signal-to-noise ratio, but it is predominantly the spectrum of the embedding medium! It shows virtually no $-C \equiv N$ absorption from the fiber, and shows intense aromatic bands from the polystyrene. The singly apertured spectrum shows a stronger ν_{CN} vibration. But the result that best represents the fiber is the third one, even though it is noisier. Since the measured area is larger, a poorly spatially resolved spectrum appears to be better. But the ν_{CN} band at 2243 cm^{-1} is attributed to the fiber only. Therefore the result where this band is strongest relative to the other bands is the most accurate result. This aperture redundancy is the heart of a high performance infrared microscope. The meaning of the word "redundant" has changed over time, and in this case what is meant is a duplicative function, rather than a useless one.

1.3 REDUNDANT APERTURING*

In order to describe the theory behind the operation of dual remote image masking, one must look at the diffraction theory of light. Specifically, the procedure is to ignore the imaging of the specimen in the microscope, and instead talk about the imaging of the sharp edged, high contrast apertures which are used to delineate the sample. Of course, the sample can also play an important role in the performance of an infrared microscope, but for now, it will be ignored. The key to Redundant Aperturing is that the first aperture is actually defining the area of the sample that is illuminated. That is, the sample is non-luminous, and the system is only "lighting-up" a small area, rather than illuminating the full field, as is done when viewing the sample. Therefore, one must know the shape of the diffraction-limited image of the aperture in the specimen plane. Since there are very few "exact solutions" in the realm of diffraction theory, it is fortunate that rectangular apertures are usually used to delineate the sample. If each blade of a rectangular aperture is considered separately, then the problem becomes that of diffraction at a high contrast straight edge. This problem has been worked out for us already [1]. Figure 3 shows the mathematical relationship between intensity and position for a high contrast edge which is defining the input of energy to the specimen.

* Redundant Aperturing™ is a trademark of Spectra-Tech, Inc.

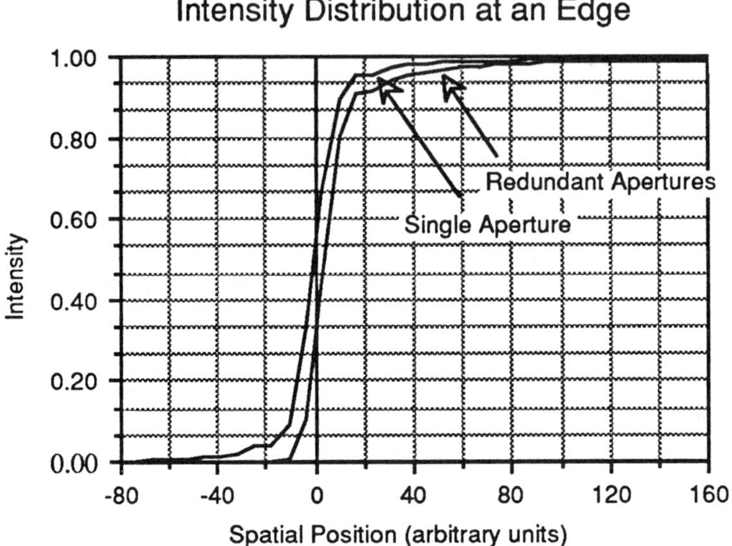

Figure 3. Intensity profile at a high contrast edge, single and dual image masking.

The ideal imaging of an edge in the absence of diffraction is a step function. Note that the energy which exists in the geometrical shadow of the edge is undesirable since it will illuminate an unwanted portion of the specimen. What is left to do is to superimpose a similar output function in the image plane after the specimen. The resultant energy profile at the detector is the product of these two functions (Figure 3). Of course further assuredness of spatial purity can be obtained by over-aperturing. This is accomplished by deliberately closing down the remote field apertures beyond the desired area of the specimen. In the size limit, however, one is often energy limited, and can not afford to throw away energy. In this situation, it is much better to carefully redundantly aperture to the edges of the area of interest. In some cases, for instance where the surrounding medium is known and can be subtracted, the second aperture may be neglected in the interest of speed.

Aperture redundancy is, however, only one of many important considerations in the design of a quantitatively and qualitatively accurate, high sensitivity instrument. Many of the other aspects were addressed in the many papers written about such instruments in the late 1940's and early 1950's [2,3,4,5].

1.4 NUMERICAL APERTURE

Chief amongst these details is the need for a high numerical aperture optical system which is also free of aberrations to below the diffraction limit, over a reasonable field. The numerical aperture of an optic is proportional to the sine of the angle of the most extreme ray accepted by the system. Typically, microscope objectives are highly corrected optical devices, employing combinations of many lenses in the design to cancel out optical aberrations such as spherical, chromatic, coma, and field curvature. Usually, field curvature is the least bothersome, since a microscope is normally used in a mode where an object of interest is brought to the center of the field of view. Of course for use in both the visible region and the mid-infrared, our choice of lens materials is extremely limited. Also, since the wavelengths of interest span a large range, the compensation for chromatic aberration becomes a more difficult problem. For these reasons and some others, it is desirable to consider the use of all-reflective optics in an FT-IR microscope.

The all-reflecting surface concept presents no major problems for the optical designer, except in the area of the objectives, and even here there is some historical information. Reflective surfaces have been *de rigueur* in the area of telescopy for many years [6]. The major concern with on-axis mirror systems is that the image falls in the path of the incoming energy, causing a partial obscuration of energy. Systems design must take this problem into account, minimizing the percentage obscuration. The result is a limitation on the number of good solutions to the design. As early as the 1900's, the all-reflective concept was applied to microscopy. Schwarzchild deduced a class of reflective microscope objective designs where both surfaces are spherical [7]. Burch [8], a number of years later, determined that for numerical apertures greater than 0.5, making one or both of the surfaces aspherical would improve performance. These designs still for some reason are colloquially called Schwarzchild configuration.

Design-wise, an objective for photometric as well as visual use differs from a visual only objective. In microscopy, the concept of "empty magnification" is used to explain the relationship between numerical aperture and magnification. For a given magnification, all available objectives from all manufacturers are virtually identical in their numerical aperture. Even though numerical aperture is proportional to the resolution of the system, resolution is also limited by the field angle subtended by the eye. Therefore after a point, the increased resolution is "empty" or useless, and increasing the numerical aperture serves only to complicate the design. Similarly, a continued increase in magnification without increased aperture results in a larger image which does not contain more detail about the specimen. In photometric applications, however, increased aperture serves to increase the energy throughput of the

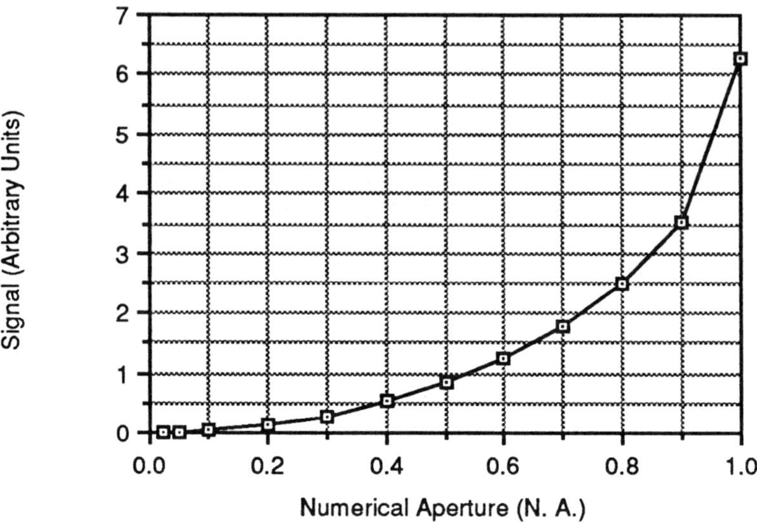

Figure 4. System performance as a function of numerical aperture.

system. This optical throughput, Θ, is directly related and proportional to the signal to noise ratio in the spectrum. Figure 4 shows the relationship between signal and numerical aperture for a microspectrophotometer system. Even for moderate magnification objectives, very high numerical apertures are desirable. A good value is N.A. = 0.5, which means that the most extreme ray is at 30 degrees from the normal. A better value is 0.71, or 45 degrees. Much beyond this, other problems occur, such as extremely small working distance, aberrations, and very small depth of field.

In summary, the numerical aperture of the objective (and condenser) of the FT-IR microscope is very important to performance, both for energy throughput and spatial resolution.

1.5 SIGNAL-TO-NOISE RATIO

Numerical aperture as it applies to the attainment of the highest possible signal-to-noise ratio result needs further discussion. This field of discussion is called radiometry, the study of the collection, imaging, and detection of radiant energy. The signal-to-noise ratio of a radiometric optical system is determined by several factors, which have been formulated into an equation by Griffiths [9] and Griffiths and de Haseth [10] for the case of an FT-IR spectrometer which is not attached to a microscope device:

$$\mathrm{SNR} = \frac{S}{N} = \frac{u_\nu(T) \cdot \Theta \cdot \Delta\nu \cdot t^{1/2} \cdot \xi}{\mathrm{NEP}} \qquad [1]$$

where S is the signal

N is the noise

$u_\nu(T)$ is the spectral energy density for a black body (derived by Planck)

Θ is the limiting optical throughput, either at the detector or interferometer

$\Delta\nu$ is the resolution

t is the measurement time

ξ is the overall system efficiency

NEP is the noise equivalent power of the detector

The spectral energy density, $u_\nu(T)$, is the quantity which describes the radiance of the source. The source in an FT-IR spectrometer is usually a nichrome wire or a silicon carbide material, both of which emit radiation which has essentially a black body profile. These sources are operated at different temperatures, depending on the manufacturer. Generally, though, the temperature is in the range of 1273 to 1573 K. The spectral energy density at a given wavelength is given by the following equation by Planck:

$$u_\nu(T) = \frac{C_1 \nu^3}{\exp\left(C_2 \nu / T\right) - 1} \qquad [2]$$

where C_1 is a constant (1.191×10^{-12} W / cm^2 sr (cm^{-1})^4)

C_2 is a constant (1.439 K cm)

ν is the frequency (cm^{-1})

T is the temperature of the black body (K)

Griffiths and de Haseth, as well as others [11,12], have found that this equation fairly accurately correlates to the measured values for FT-IR instrumentation. This would indicate that the current hardware is working very near to its theoretical limit. This is not to say that the signal to noise will not improve in the future. This can happen through improvements to detectors, sources, or system efficiency.

For the purpose of predicting the theoretical performance of an FT-IR microscope, two changes to this equation are needed. First, in the microscope, neither the detector nor the aperture usually limits the optical throughput. Instead, the specimen limits throughput in a general purpose FT-IR microscope. One usually operates the FT-IR microscope in a mode where the specimen is brought to the center of the field of view, and then delineated by the variable apertures situated in remote image planes of the specimen. Normally, this specimen plane is imaged onto the detector, so that is is possible to customize the detector size for specific applications, and in that manner increase performance. This means that the throughput of the system changes, based on the size of the specimen. There are two important consequences of this.

First, in the diffraction-limited region, the energy through the system is governed by the aperture size as well as the source profile. Therefore, it is imperative to measure both sample and background spectra through the same aperture size, else a sloped baseline will occur. The best way to avoid this is to measure the sample spectrum immediately after aperturing down on the specimen, then moving the specimen out of the way to record the background through the same aperture. Fortunately, this is the most convenient way to run a microscope anyway, allowing excellent compensation of water vapor, and minimizing problems from long term instrument drift.

The second consequence is that the system can only be throughput-matched for one specimen size, and for smaller or larger specimens, the performance will not be ideal. The normal procedure is to throughput-match for a very small specimen. When a large specimen needs to be measured, the system is not optimized, but there is plenty of energy anyway.

The throughput at the specimen is the product of the specimen area (A), and the solid angle subtended by the objective from the specimen (Ω_s):

$$\Theta_s = A\Omega_s \text{ cm}^2\text{-sr} \qquad [3]$$

Let us assume that the specimen is self-luminous. The specimen can be thought of as being the source of the system. The solid angle is the portion of a sphere that a source of optical radiation radiates into when that source is at the center of the sphere. If the source radiates into the full sphere, this is equivalent to 4π steradians (sr). The solid angle may be evaluated in terms of the surface area of the portion of the sphere that is radiated into (S) and the radius of the sphere (R):

$$\Omega_S = \frac{S}{R^2} \qquad [4]$$

It would be more convenient for our purpose, however, to express the solid angle in terms of the limiting numerical aperture of the microscope (usually the N.A. of the objective). Let us consider a sphere of unit radius, depicted here in two dimensions (Figure 5).

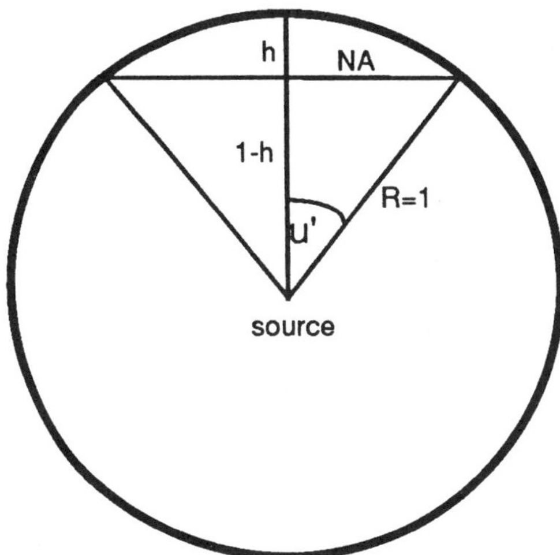

Figure 5. The construction of a unit radius sphere used to calculate solid angle in terms of the numerical aperture (see text).

A source is placed at the center of the sphere, where it radiates into a u' semi-field angle, and therefore a 2u' included angle. The quantity h is used in an equation which gives the surface area of the sphere irradiated by the source:

$$S = 2\pi Rh \qquad [5]$$

Note that for the unit radius sphere, the value of the numerical aperture forms a side of a right triangle with a hypotenuse of unity, and the other side equal to (1-h). This allows us to express h in terms of the numerical aperture by the Pythagorean relation:

$$(1-h)^2 + (N.A.)^2 = 1 \qquad [6]$$

Rearranging:

$$h = 1 - \sqrt{1 - (N.A.)^2} \qquad [7]$$

Therefore Eqn. 8 can be written:

$$\Omega_S = S = 2\pi \{ 1 - \sqrt{1 - (N.\ A.)^2} \} \quad [8]$$

Next, a factor must be brought into the equation to predict the losses associated with the diffraction effect, because of the small sizes of specimens being observed. Smith [13] puts forth an equation which relates the irradiance at the center of the Airy disk pattern (H_0) to the total power in the Airy disk (P):

$$H_0 = 3.5\ P\ (N.A. \div \lambda)^2 \quad [9]$$

where λ is the wavelength. This equation predicts an efficiency due to diffraction of 1% for our system when the object being measured is small compared with the wavelength of light. The FT-IR microscope, however, is used also for larger samples where the diffraction effect still lowers system efficiency, but not to the extent predicted by Smith's equation.

Rather than developing the general equation for the efficiency due to diffraction for an FT-IR microscope, a graph of efficiency versus sample size is developed for rectangular samples in the IR-plan microscope with the 15x magnification, N. A.=0.58 objective, at 2000 cm^{-1}. Under incoherent illumination, the imaging of a high-contrast edge, such as the blades of the apertures used to delineate the sample in an FT-IR microscope, is described by the function:

$$I_{im}(x) = \int_0^\infty I_{ob}(\xi)\ \text{sinc}^2 x\ \ d\xi \quad [10]$$

where I_{im} and I_{ob} are the intensity at the image and the object, respectively. The graph of this function for two parallel edges, as would be used to delineate a specimen, is shown in Figure 6. It is seen that the energy falling outside the rectangular area of the graph is lost due to diffraction. This energy is imaged outside the desired area, and therefore manifests itself as stray light. If redundant aperturing is used, much of it is effectively blocked by a second aperture. In either case, since this energy does not go through the part of the specimen that one wishes to sample, one should consider it as reducing the system efficiency. As the size of the specimen approaches the wavelength of light, this efficiency approaches zero. Figure 7, the graph of efficiency due to diffraction, can be constructed based on the percentage diffracted area demonstrated in the preceding graph.

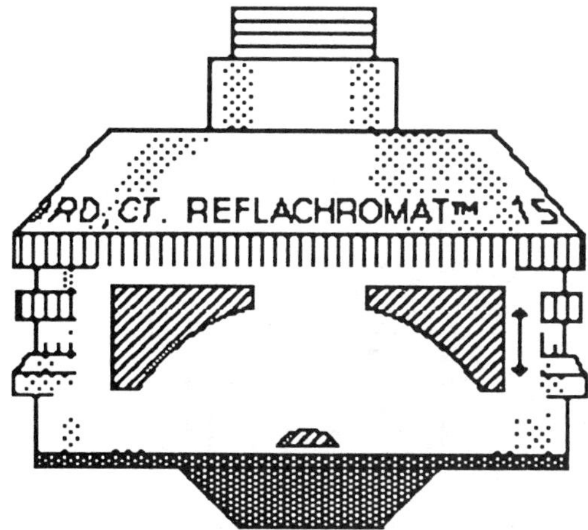

Figure 8. The optical configuration of the Schwarzchild-type Cassegrainian objective lens. The large primary mirror moves down to compensate for spherical aberrations.

1.7 COHERENCE

The need to discuss the coherence of the light being used in the FT-IR microscope arises from the fact that image quality is affected by coherence, in situations where the diffraction limit is approached. This was realized as early as 1893 by Abbe [15], and illustrated experimentally in 1906 by Porter [16]. Rather than delving into the mathematical basis for this effect, an example would be in order. In a microscope measurement, the light is always at least partially coherent. That is, energy from neighboring points in the object (source), because of diffraction, tends to overlap and correlate. The visual result is that fringes (light and dark alternating features) are formed in the image plane of the microscope. These are considered undesirable from the standpoint of visual image quality, and effort is made in a microscope system to minimize these features. One way to lessen the degree of coherence in a microscope is to place a diffusing glass after the source, although this is not extremely useful as a scrambler. Rotating diffusers are better, or even liquid diffusers, like milk solutions have been used. Liquid diffusers work because of Brownian motion in the milk suspension [17]. Time-averaged observation can also break down coherence, because of Brownian motion in the atmosphere. The degree of coherence is also related to the numerical aperture of the optical

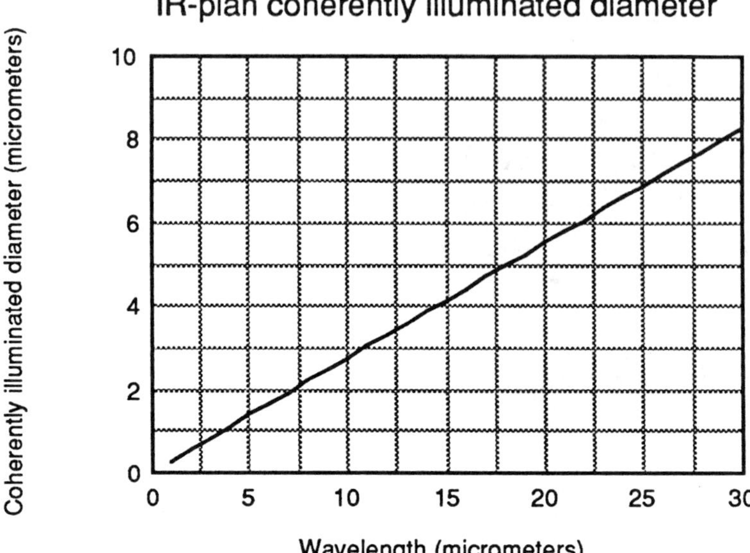

Figure 9. Coherence as it relates to the FT-IR microscope. For very small specimens, the system is almost totally coherent.

system and the mean wavelength of illumination [18]. The Smith-Helmholtz theorem gives rise to an equation (Eqn. 12) which expresses the diameter of the coherently illuminated area of the exit pupil of a system (d'_{coh}) illuminated by an incoherent quasi-monochromatic uniform circular source:

$$d'_{coh} \sim (0.16\,\lambda_0) \div (\text{N. A.}) \qquad [12]$$

where λ_0 is the mean wavelength and N.A. is the numerical aperture. Figure 9 evaluates this relationship for the IR-plan FT-IR microscope operating in the mid-infrared region. This shows that when dealing with objects at the limit of detection, that is in the 5 micrometer diameter region, the illumination in the IR-plan microscope is fairly coherent. For larger specimens, the coherence interval falls off.

For the FT-IR microscope, one is again faced with a dichotomy. While incoherence is desired for the viewing step, an increased degree of coherence would give rise to a sharper edge function, slightly improving spatial resolution. Unfortunately, while there are several ways to destroy coherence, the only way to increase it is to lower the effective angular subtense of the optical system. Given the constraints of the type of measurement being made, this is only possible by decreasing the numerical aperture of the system. The small benefit imparted would be more than offset by the increase in the size of the Airy disk, and would also be undesirable from an energy efficiency standpoint.

The temporal coherence of the illuminating energy does not affect the edge function unless chromaticity exists in the optical system [19]. The implication here is that if our optical system has residual uncorrected chromatic aberration, then an improved edge function for a narrower bandwidth source will be realized. Since most FT-IR microscopes are totally reflecting systems, chromatic aberration is not an issue, mirrors being totally achromatic.

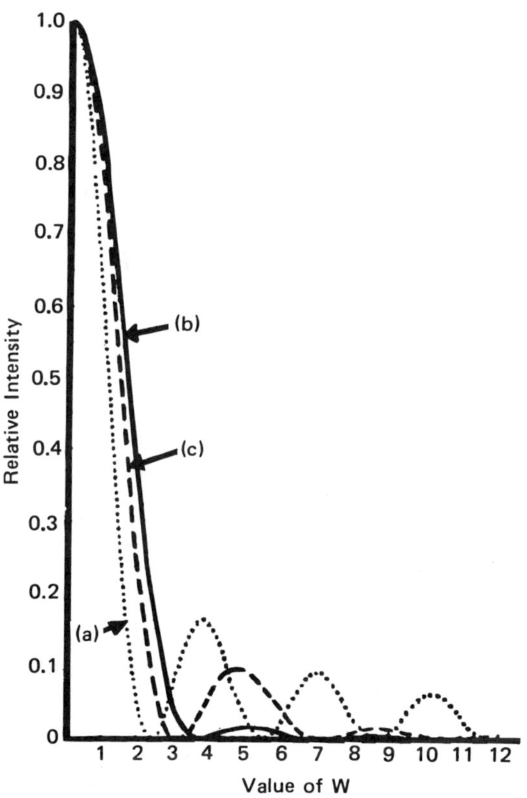

Figure 10. The Airy disk diffraction pattern is modified when the optical system contains a central obscuration, as is the case in most FT-IR microscopes. This figure shows the modification for various levels of central obstruction.

1.8 PUPIL APODIZATION

A by-product of our need for a totally reflecting optical system for the FT-IR microscope is that the diffraction limited, Cassegrainian optics that must be used have a central obscuration (see figure 8) which affects the diffraction imagery of the system. Figure 10 shows that the net effect of this central obscuration is to modify the Airy disk diffraction pattern, increasing the intensity in the first few lobes, and reducing the energy in the central bright maximum, while at the same time narrowing it. The Airy disk, and it's meaning to the imaging of a microscope system has been described previously, and the reader is referred here for more information [20].

1.9 CONCLUSIONS

The engineering aspects of the modern infrared microscope/spectrometer have been overviewed here, in order that the reader may understand the apparatus, and use its full potential. Redundant Aperturing and Sample Compensation combine to improve the spatial resolution capabilities of the FT-IR Microscope to the point where the wavelength of light is truly the limiter of further gain. Infrared microspectroscopy now stands as the best choice for identification of small amounts of organic materials, and for the measurement of or correlation with a number of their physical properties such as dichroic ratio or physical strength.

ACKNOWLEDGMENTS

I would like to extend thanks to Professor Peter Griffiths of the University of California - Riverside for helpful discussions regarding the signal-to-noise equation, to Dr. Bruce Chase of DuPont for discussions about beamsplitter efficiency, to Provost Brian Thompson of the University of Rochester for insight into the significance of coherence in microscope systems, to Dr. Lola Darmon of Eastman Kodak for the embedded fiber experiment, and to Dr. John Reffner of Spectra-Tech for various helpful discussions.

REFERENCES

1. B. J. Thompson, in *Progress in Optics* (E. Wolf, ed.), Vol. 7, Elsevier North-Holland, New York (1969).

2. V. J. Coates, A. Offner, and E. H. Siegler, Jr., *J. Opt. Soc. Am.*, 43: 984-989 (1953).
3. R. Barer, A. R. H. Cole, and H. W. Thompson, *Nature*, 163: 198 (1949).
4. R. C. Gore, *Science*, 110: 710 (1949).
5. E. R. Blout, G. R. Bird, and D. S. Grey, *J. Opt. Soc. Amer.*, 40: 304 (1950).
6. A. Bouwers, *Achievements in Optics*, Elsevier, New York (1950).
7. K. Schwarzchild, *Astr. Mitt. Königl. Sternwarte Göttingen*, Göttingen, (1905).
8. C. R. Burch, *Proc. Phys. Soc.*, 59: 41 (1947).
9. P. R. Griffiths, *Chemical Infrared Fourier Tranform Spectroscopy*, Wiley, New York (1975).
10. P. R. Griffiths and J. A. de Haseth, *Fourier Transform Infrared Spectrometry*, Wiley, New York (1986).
11. C. T. Foskett and T. Hirschfeld, *Appl. Spectrosc.*, 31: 239 (1977).
12. D. R. Mattson, *Appl. Spectrosc.*, 32: 335 (1978).
13. W. J. Smith, *Modern Optical Engineering*, McGraw-Hill, New York (1966).
14. R. G. Messerschmidt, in *The Design, Sample Handling, and Applications of Infrared Microscopes* (P. B. Roush, ed.), ASTM STP 949, American Society for Testing and Materials, Philadelphia (1987).
15. E. Abbe, *Arch. Mikroskop. Anat.*, 9: 413 (1893).
16. A. B. Porter, *Phil. Mag.*, 11: 154 (1906).
17. B. J. Thompson, *op. cit., reference 1.*
18. M. Born and E. Wolf, *Principles of Optics*, 6th Ed., Pergamon, Oxford (1980).
19. B. J. Thompson, *personal communication.*
20. R. G. Messerschmidt, *op. cit., reference 14.*

2

Infrared Microspectroscopy: Development and Applications of Imaging Capabilities

MATTHEW A. HARTHCOCK AND SCOTT C. ATKIN *Analytical and Engineering Sciences, Texas Applied Science and Technology Laboratories, Dow Chemical U.S.A., Freeport, Texas*

2.1 INTRODUCTION

Considerable interest has recently been shown using infrared microspectroscopy or infrared microprobe spectroscopy for characterization of microscopic materials, defects, etc. (for example, see references 1-4). Additionally, the microscopic characterization or compositional mapping of materials has been done using a variety of imaging technologies. The technologies range from elemental imaging techniques such as electron microprobe [5,6] to techniques such as magnetic resonance imaging [7]. This by no means covers all of the imaging techniques. Several new technologies are being developed that allow images of a material to be obtained based on functional group maps. These include surface enhanced Raman spectroscopy (SERS) [8], coherent anti-Stokes Raman spectroscopy (CARS) [9], and laser electron microscopy [9]. Presented in this chapter is the research that is being conducted to develop infrared microspectroscopic functional group imaging.

Applications of imaging techniques encompass all fields of science including, for example, polymer/materials science, the biological/medical sciences, and the chemical sciences. Examples of infrared microspectroscopic functional group imaging in the polymer science field will be discussed in order to demonstrate the capabilities of the technique and its advantages over "standard" infrared microspectroscopic analyses.

2.2 EXPERIMENTAL

2.2.1 Basic instrumentation

Figure 1 shows a schematic of the configuration of the instrumentation used for infrared microspectroscopic functional group imaging analysis. A Digilab (Cambridge, MA) FTS-50 Fourier transform infrared spectrophotometer equipped with a Digilab universal transmittance/reflectance microscope (UMA-100) and a narrow band MCT detector was used for collection of the infrared spectra. The infrared spectra were recorded using a Cassegrainian objective with an optical magnification of 32X. The sample stage is a Digilab computer controlled x,y mapping stage with a controller capable of controlling the steps of the stage to as small as 10 micrometers. Automated infrared spectral data collection was done by controlling the stage via the controller using the 3200 data station of the FTS-50 spectrophotometer. The stage can be moved in a one or two dimensional direction and infrared spectra recorded at the various desired spatial coordinates. Details of the experiments will be given later. A high speed personal computer was interfaced with the 3200 data station of the FTS-50 instrument for reduction of the spatially specific infrared spectroscopic data into images based on functional group absorption maps.

2.2.2 Experimental details

This section will describe the procedure for obtaining a functional group image, one and/or two-dimensional, of a material. Several parameters are necessary to establish the conditions of a functional group image, they are: the step size in the x and y direction (defined as the plane in which the sample resides, the z direction would correspond to the pathlength), the number of steps to be taken in the x and y directions, the aperture size through which the infrared radiation will be allowed to pass and be detected (250 micrometers or less as defined by the detector size), and the absorption frequency at which the functional group image is desired.

Figure 2 shows the definition of the parameters necessary to perform an actual mapping experiment. The small box is the aperture above the sample (see Figure 1) with the size denoted a_x by a_y (shown here to be a square but rectangular and circular apertures can also be used) and s_0, s_1, etc. denote the step size where the subscript represents the number in the x direction. This particular example is representative of a one dimensional map.

Once the infrared spectra have been collected through the given aperture and as a function of the x and/or y spatial dimensions, the spectra are reduced to an intensity measurement (denoted I_0, I_1, etc. in Figure 2 and can be absolute absorbance, relative absorbance, etc.) at a given frequency which corresponds

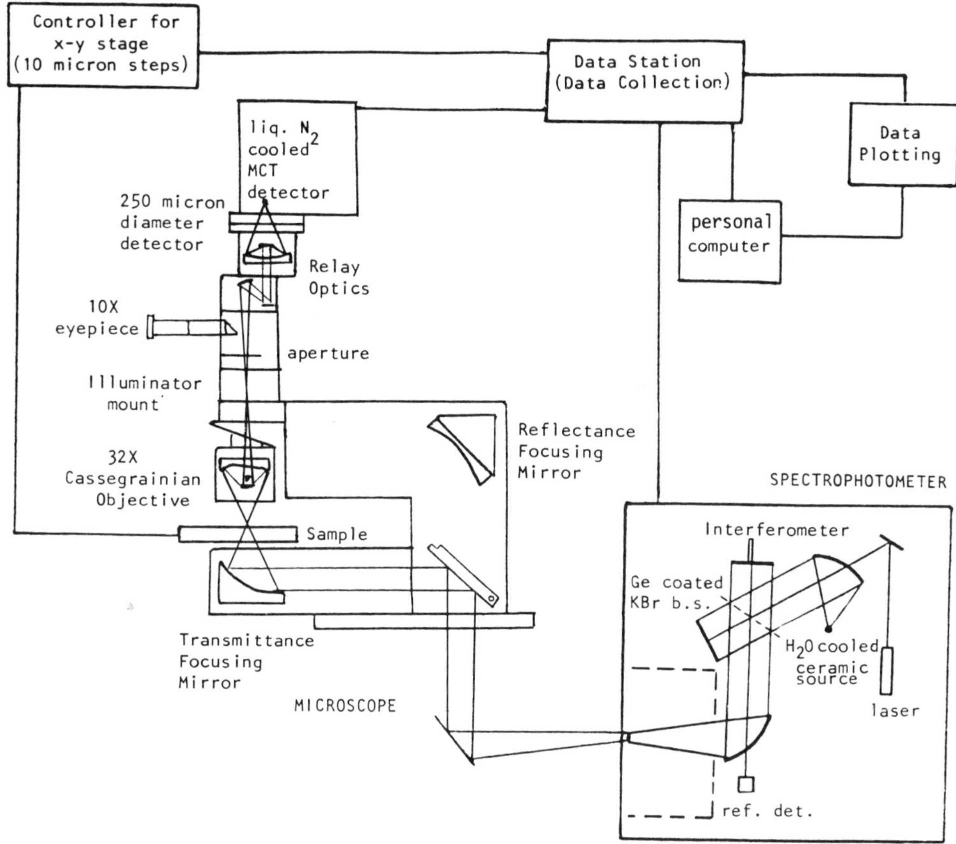

Figure 1. Schematic of the configuration of the instrumentation used for infrared microspectroscopy functional group imaging.

to a particular functional group absorption of infrared radiation. Once the absorbance at a given frequency is obtained as a function of the x and y spatial coordinates, an image can be obtained. As observed in the figure, the intensity values (absorbance, etc.) are linearly interpolated for graphic representation. Also, when representing the image as a gray level image or a contour map, to be described later, the number of levels can be varied using the data experimentally obtained in order to vary the amount of detail in the image and to better quantitate differences in the composition of a particular functional group in the material as a function of the spatial coordinates. The range, R, is the difference between the quantitatively determined high and low absorbance (intensity) values. This range can be subdivided into increments that effect the detail of the functional group images.

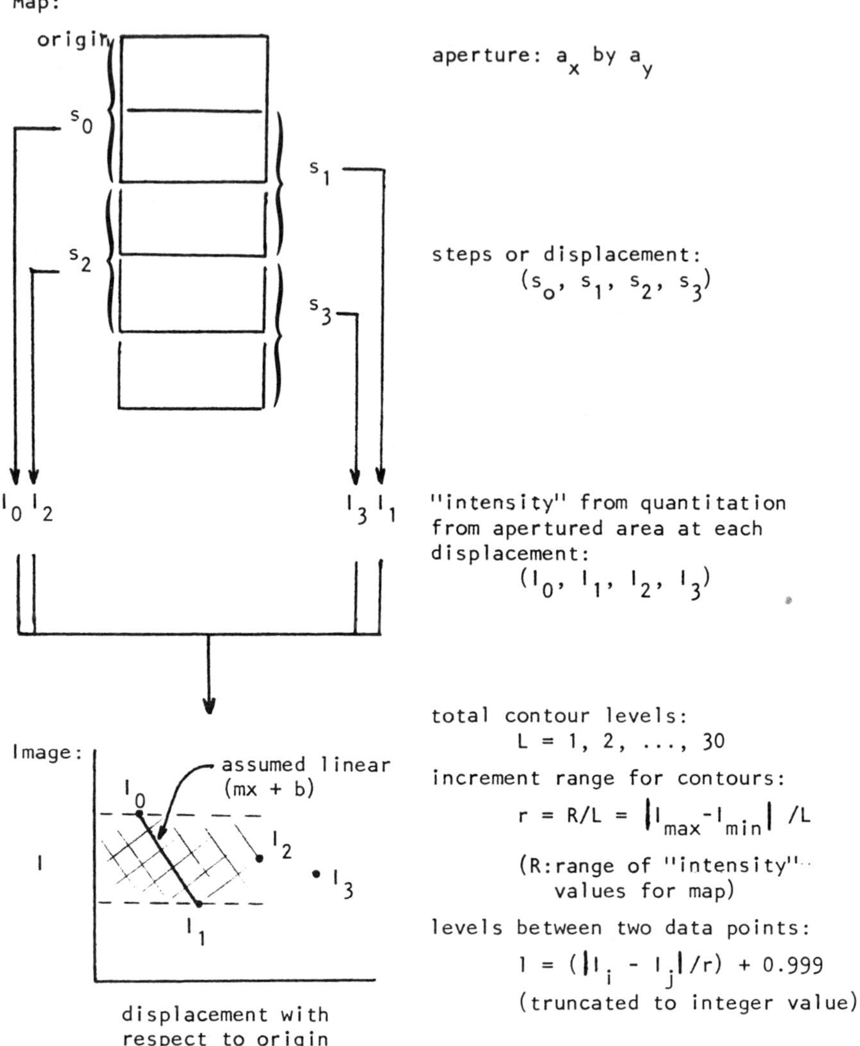

Figure 2. Definition of terms used for collecting infrared spectra as a function of specific spatial coordinates which are necessary for obtaining functional group images.

The one-dimensional map (image) is represented using a simple plot of intensity versus coordinate value from a reference point. The two-dimensional images can be represented by: false color maps, gray level maps, axonometric plots, and contour plots. Computer software to quantitatively reduce the data and to represent the image as a false color plot (represented as gray level images) has been developed in our laboratory for the applications described here. The axonometric plots (three dimensional type plots) complement the false color and contour plots. The desired form of representation is dependent on the analysis required from the data and/or representation which is easiest to interpret.

When considering the use of rectangular apertures, three types of experiments can be performed with respect to the step and aperture sizes. Table 1 summarizes the various cases and lists comments related to each case. In the case where the step size and aperture dimension are equal, theoretically no overlap of adjacent areas would occur and the spatial resolution would be optimized to the point where two adjacent areas of a material could be distinguished. However, Messerschmidt, in chapter 1 of the volume, pointed out that in practice this would not occur. For the case in which the step size is larger than the aperture dimensions (case II), spatial domains would not be accurately represented because complete sampling would not occur. When the step size is less than the aperture dimensions (case III), spectra would not necessarily be unique and the spatial resolution for distinguishing differences between two adjacent non-similar areas could be improved. The resolution improvement can be done using subtraction techniques and other mathematically assisted methods.

Case III in Table 1 is the condition which is most used for performing a map. This allows improvement in the effective spatial resolution for distinguishing adjacent regions and for insuring representative sampling of the material. The diffraction of infrared radiation when using apertures of approximately 10 micrometers or smaller does not immediately lend to the spatial resolution being improved using the experimental design described here [10].

To be noted is that numerous functional group images can be generated since the "complete" infrared spectrum (in our case limited by the use of a narrow band MCT detector) can be used for reduction to specific functional group images.

2.3 RESULTS AND DISCUSSION

The use of functional group images obtained using infrared microspectroscopy has resulted in a technique that can be used to investigate hypotheses that could

Table 1. Various cases for imaging experiments based on step and aperture size.

CASE	CONDITION	COMMENTS
I	step size = aperture dimension	-no overlapping areas -spatial resolution optimized
II	step size > aperture dimension	-spatial domains not accurately represented
III	step size < aperture dimension	-overlapping areas -spatial resolution can be improved by a factor of approximately two

not previously be addressed. There are several reasons why the technique enhances and improves the analytical use of infrared microspectroscopy. First, frequently when a sample is prepared for analysis (for example by microtoming) the material to be analyzed by infrared microspectroscopy may no longer be visible using standard optical microscopy techniques and/or accessories. Thus, a two dimensional map of the material may be made and various functional group maps can be generated to obtain an image of the material. The image which results is a spatially resolved determination of the chemical composition.

Second, generating an image of the material allows the material to be representatively sampled. This is particularly important since often the sample can be considered to be macroscopic as defined by the microprobe technique. For example, if an imperfection in a material is one millimeter in size, this is several times larger than the detector (250 micrometer diameter) and representative sampling would provide confidence in the results of the analysis.

Finally, the technique allows automated sampling of a material. This reduces the amount of "hands-on" time the spectroscopist must spend with a given analysis. Each of the three enhancements that functional group imaging makes to infrared microspectroscopy will be demonstrated using the examples which follow.

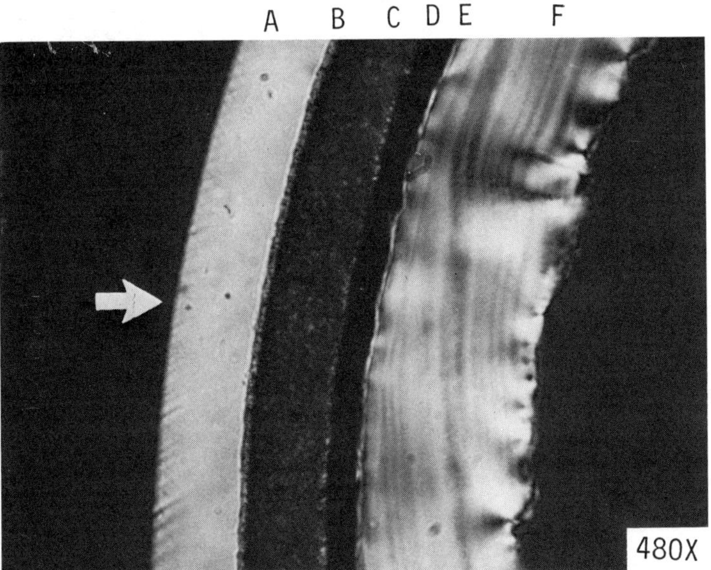

Figure 3. Optical micrograph of a microtomed cross-section of a multilayer polymer structure (480X magnification). Experimental conditions for collecting spatially specific infrared data: (1) 15 x 75 micrometer aperture, (2) 10 micrometer step size, (3) 256 scans for signal averaging, (4) 8 cm $^{-1}$ resolution, and (5) transmittance spectra.

2.3.1 Multilayer Polymer Structure (One-dimensional Functional Group Image)

Figure 3 shows the optical micrograph of a microtomed cross-section of a multilayer polymer structure prepared for analysis by infrared micro-spectroscopy using a one-dimensional map. The experimental data collection parameters were: aperture dimensions 15 x 75 micrometers, step size 10 micrometers, and 256 scans collected for signal averaging at 8 cm^{-1} resolution. As can be observed from the optical micrograph, the section consists of several layers while the entire thickness of the structure is only approximately 110 micrometers. Twelve spectra were collected across the section using the computer controlled mapping stage and the conditions mentioned above.

Several functional groups were selected for mapping based on knowledge of the composition of multilayer polymer structures. The CH_3 deformation at 1374 cm^{-1} was selected to map the presence of methyl groups that could be due

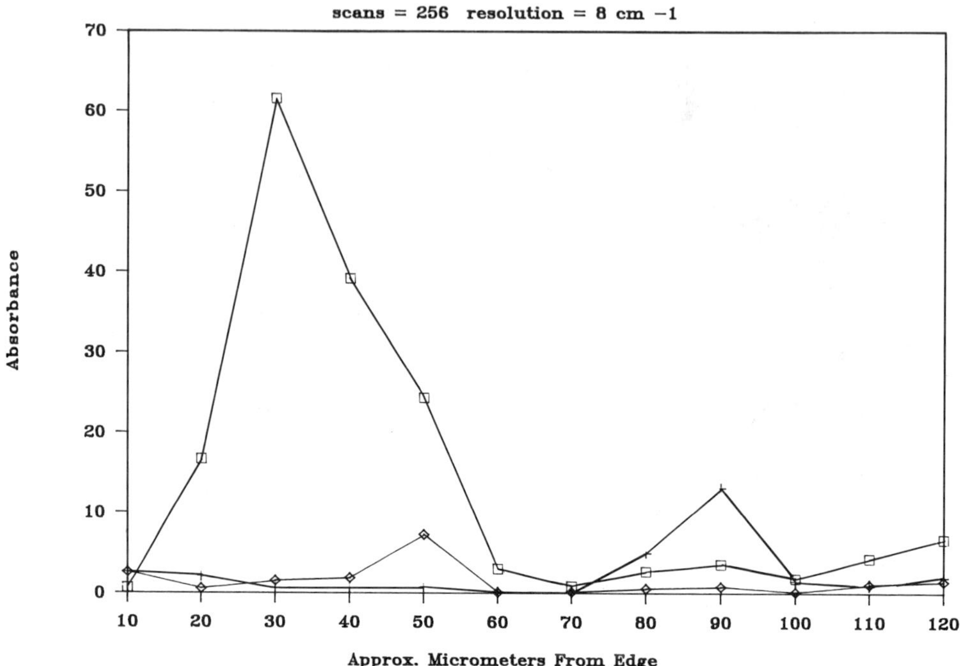

Figure 4. Functional group image of the multilayer polymer structure shown in Figure 3. Functional group images for the 1700 cm^{-1} (+), 1648 cm^{-1}(◇), and 1374 cm^{-1} (□) absorptions are given.

to polypropylene or various branched polyethylenes. The 1700 cm^{-1} absorption was used to map the concentration of acid groups that could be present in materials such as ethylene/acrylic or methacrylic acid copolymers used as adhesives. An absorption at 1646 cm^{-1} was selected to look for other types of adhesive material. Figure 4 shows the functional group map at each of these frequencies. The map indicates that at least five distinct layers are present. From the optical micrograph five and possibly six layers are present, one of which is aluminum foil (layer D). The layer associated with the 1646 cm^{-1} absorption is labelled layer B in the optical micrograph in Figure 3.

From the map in Figure 4, infrared spectra can be identified from the map that are representative of the various layers. The spectra are shown in Figure 5. Layer A is identified as polypropylene. Layer B is only 2-3 micrometers in thickness. The spectrum of layer B is actually a composite of layers A, B and C because of the spatial resolving power limitations. The spectrum shows the presence of absorption bands at 1735, 1646, 1271, and 1034 cm^{-1}. These bands are likely due to urethane (or imide) type structures used for adhesion. These particular absorption bands were not observed until the mapping experiment

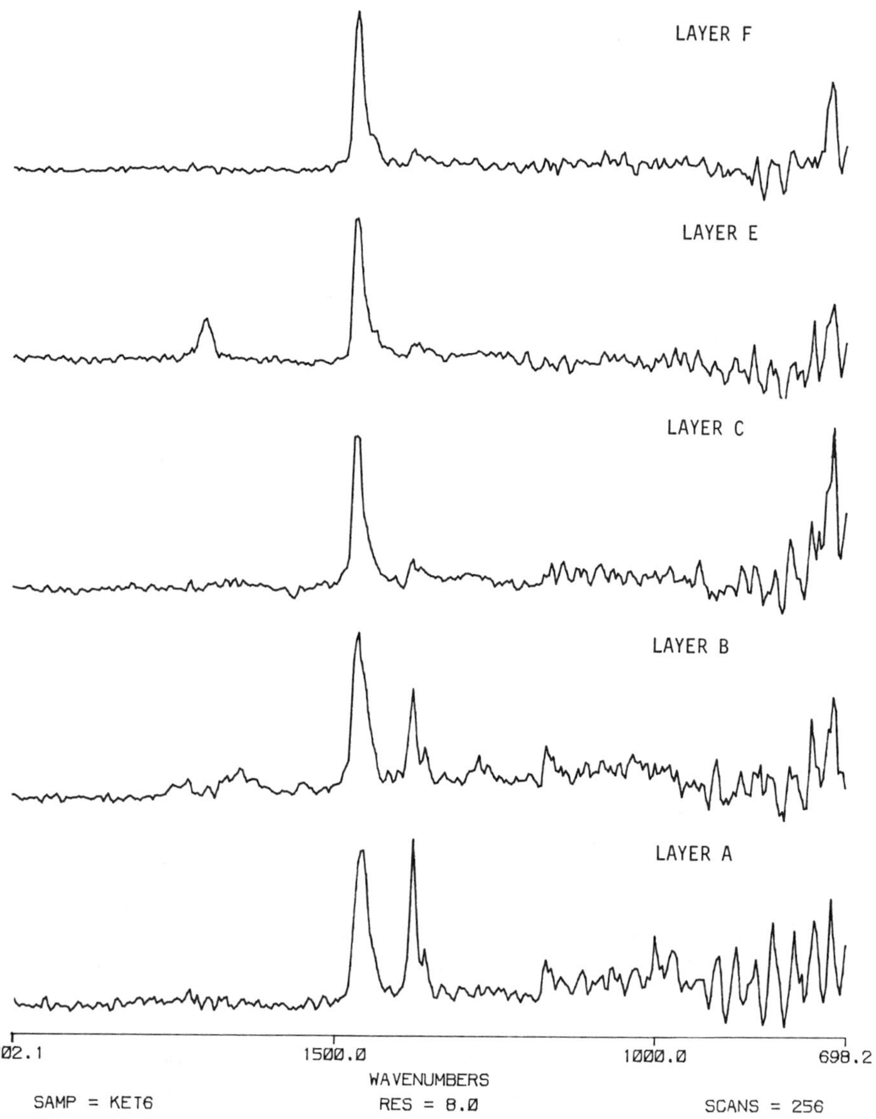

Figure 5. Representative and complete infrared spectra for the various layers of the multilayer polymer structure shown in Figure 3. (Layer D is aluminum foil and no spectrum was thus obtained for this layer).

was performed, nor was the layer conclusively determined to be present. This demonstrates how the spatial resolution for observing adjacent regions can be improved to a resolution of effectively 2-3 micrometers. The spatial resolution with which infrared spectra can be obtained is dependent on the material and conditions used to record and manipulate the infrared spectra.

Layer C is identified as polyethylene. Layer D is the aluminum foil layer. Layer E is identified as an ethylene/acrylic acid copolymer and Layer F is identified as a polyethylene type material.

From such a map, the size of the various layers can be determined to within the following conditions

$$((N-1)\cdot s) - a < X < ((N\cdot s) - a) \qquad \text{for } N > a/s$$

where X is the size of interest, N is the number of spectra showing a specific functional group absorption, a is the aperture size and s is the step size. This information can be used to complement the optical microscopy data and to determine more accurately the spectral correspondence to areas observed in the optical micrographs.

2.3.2 Imperfection in Polyethylene Film (Two-dimensional Functional Group Image)

As mentioned previously, advantages of the imaging technique are representative sampling and attainment of a functional group image of a material for problem solving. Figure 6 shows the optical micrograph of an imperfection (inclusion or gel) in a polyethylene film. In this case the imperfection visually appears to be approximately 190 micrometers in diameter. A two dimensional map of the sample was obtained using 30 micrometer steps of the mapping stage and an aperture size of a 50 micrometer square (32 scans and 8 cm^{-1} resolution were used to record the infrared spectra). The specimen was studied without any preparation.

The two dimensional array of infrared spectra was reduced to a functional group image based on the 1738 cm^{-1} carbonyl absorption in order to search for the presence of oxidation in the imperfection. Figure 7 shows the gray level image of the sample and it is obvious that a higher concentration of carbonyl functionality material exists in the imperfection than the surrounding polyethylene film. Analysis of the infrared spectra from the imperfection and the surrounding area showed a definite carbonyl absorption characteristic of oxidized polymer.

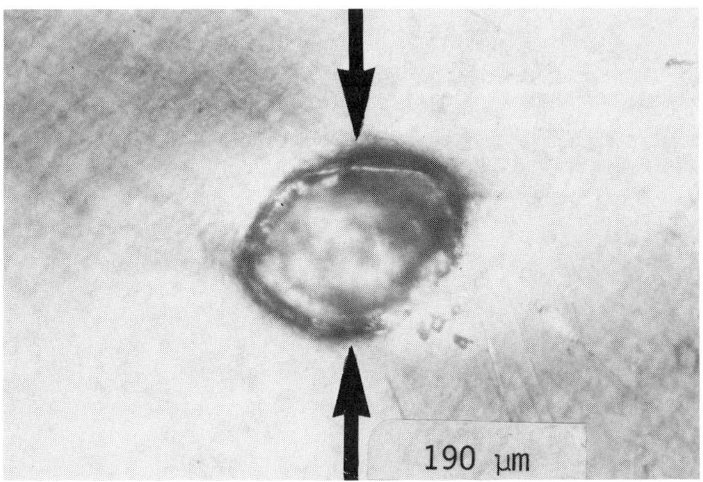

Figure 6. Optical micrograph of an imperfection in a polyethylene film (120X magnification). Experimental conditions for collecting spatially specific infrared spectra: (1) 50 x 50 micrometer aperture, (2) 30 micrometer step size in the x and y direction, (3) 32 scans for signal averaging, (4) 8 cm^{-1} resolution, and (5) transmittance spectra.

As mentioned in the experimental section, the data may be represented in several ways. Figure 8 shows the functional group image as axonometric plots and a contour plot. The axonometric plots show two formats each of which indicates the higher concentration of carbonyl groups in the imperfection than in the surrounding film. This example shows how representative sampling can be obtained and problem solving accomplished using a functional group image of the sample.

2.3.3 Surface Imperfection (Two-dimensional Functional Group Image)

Not only can the imaging be done using infrared absorption spectra, but reflectance spectra can also be used to obtain an image of the surface of a material. To demonstrate this and how an image can be obtained which complements visual observations using a chemical composition map, an example which was previously published will be used [2]. The example is that of imperfections or “fisheyes” that are present on the surface of a polyurethane material after being painted.

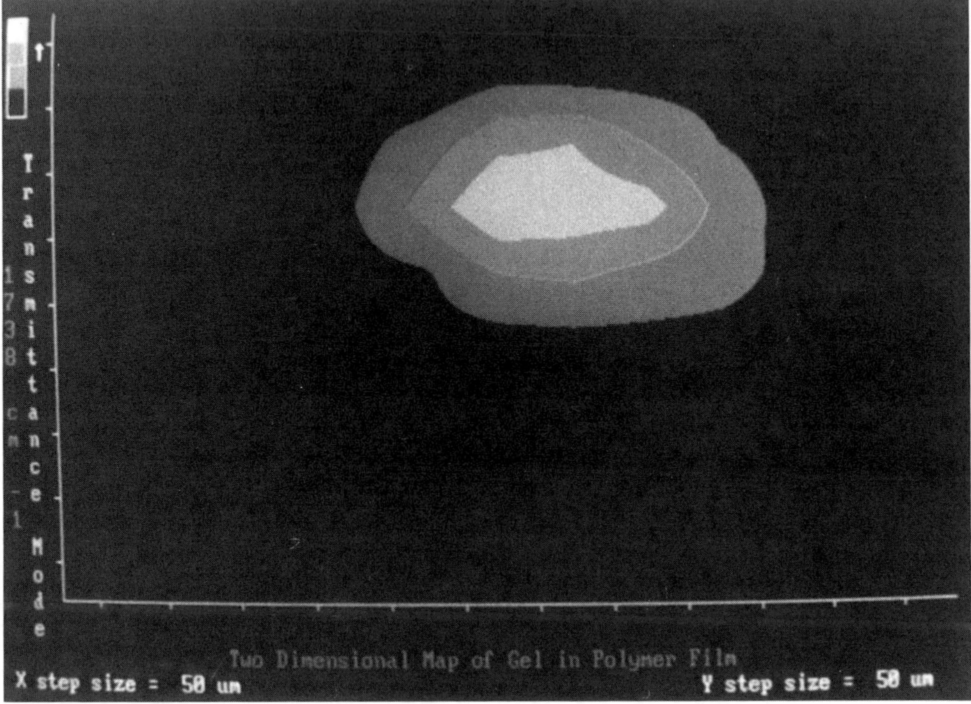

Figure 7. Functional group image of the imperfection in a polyethylene film shown in Figure 6. The image is based on absorptions at 1738 cm^{-1} and is represented as a gray level image. The number of contour levels, L, as defined in Figure 2 is four.

Figure 9 shows the optical micrograph of the surface of the polyurethane part with the imperfections. The mapping conditions used for recording the spectra were: 100 micrometer square aperture size, 20 micrometer step sizes in the x and y direction, 256 scans for signal averaging, and 8 cm^{-1} resolution. Figure 10 shows the infrared reflectance spectra previously published representative of: (1) the surface imperfection, (2) the paint on the surface, (3) the polyurethane material, and (4) mold release that was the contributing factor to the surface imperfections [2]. From the spectra, it is obvious that the mold release is a straight chain hydrocarbon and that some of the material remained on the polyurethane surface before painting, leading to surface imperfections. From the carbon hydrogen stretching vibrations at 2920 and 2852 cm^{-1} a

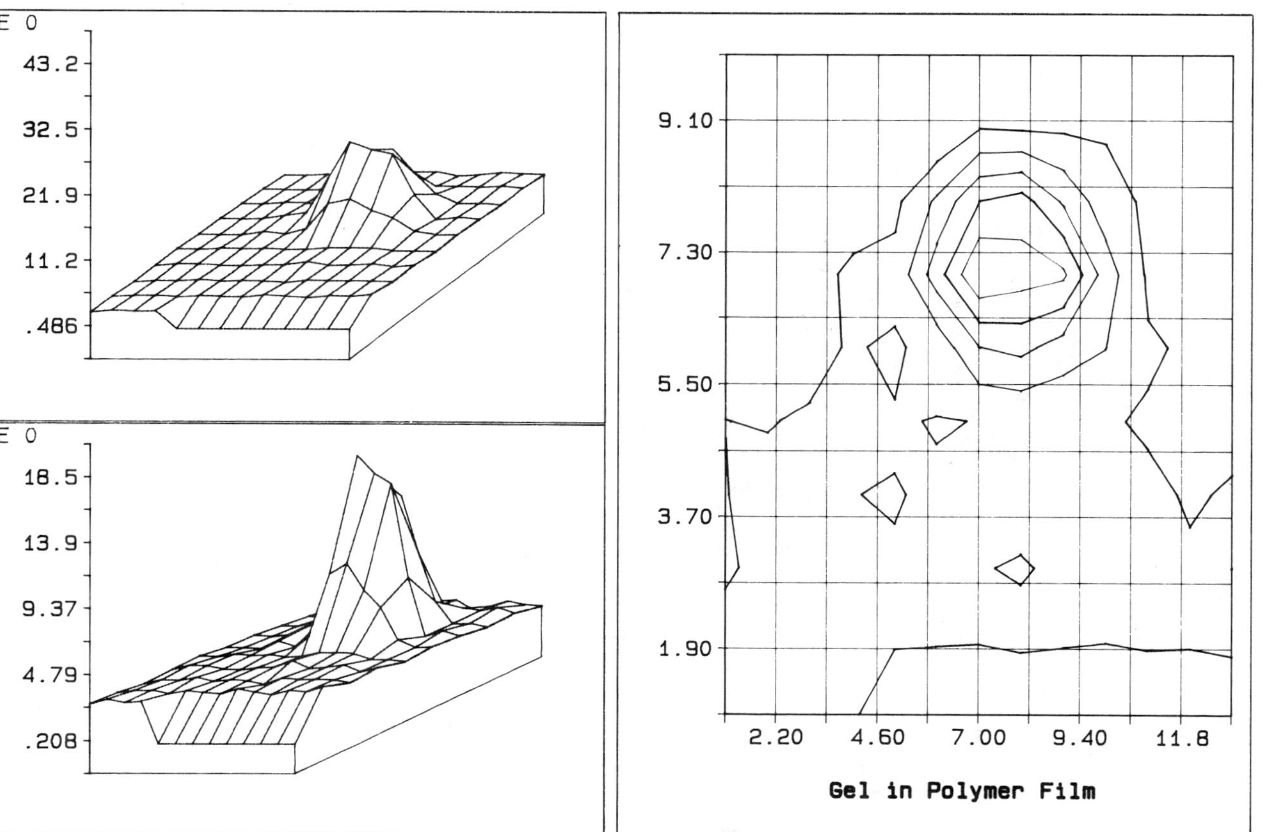

Figure 8. Functional group image of the imperfection in a polyethylene film shown in Figure 6. The image is based on absorptions at 1738 cm^{-1} and is represented by both axonometric and contour plots.

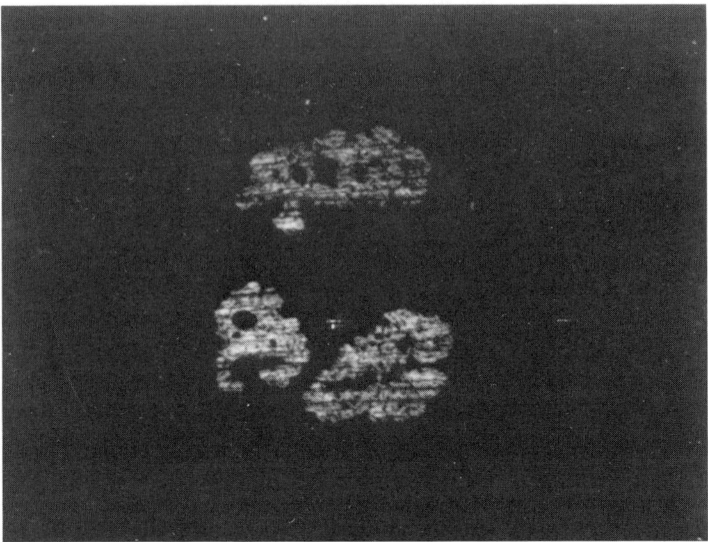

Figure 9. Optical micrograph of a surface imperfection of a polyurethane material (60X magnification). Experimental conditions for collecting spatially specific infrared spectra: (1) 100 x 100 micrometer aperture, (2) 20 micrometer step size in the x and y direction, (3) 256 scans for signal averaging, (4) 8 cm^{-1} resolution, and (5) reflectance spectra.

functional group image can be constructed of the surface. Figure 11 shows gray level images of the surface imperfection at various levels of detail. Figure 11a shows the image in which two levels (for definition of the subdivision of the range refer to Figure 2) were used. From the image, only two of the non-painted surface areas observed in the optical micrograph are observed. As the number of subdivisions are increased (Figures 11b and 11c), the detail of the functional group and its correspondence to the optical micrograph is improved. In Figure 11d the painted areas in the interior of the large imperfection can even be identified. Figure 11e shows the image resulting from subdividing the range into thirty levels. Little new information or detail is obtained from this image versus the one shown in Figure 11d.

The image is also represented as an axonometric plot in Figure 12. Three surface imperfections are evident in this representation of the data.

In general, the significance of quantitative differences in the various regions of the map based on the collection conditions (aperture size, step size, etc.) have not been completed. This information is necessary to better interpret and utilize the capabilities described here.

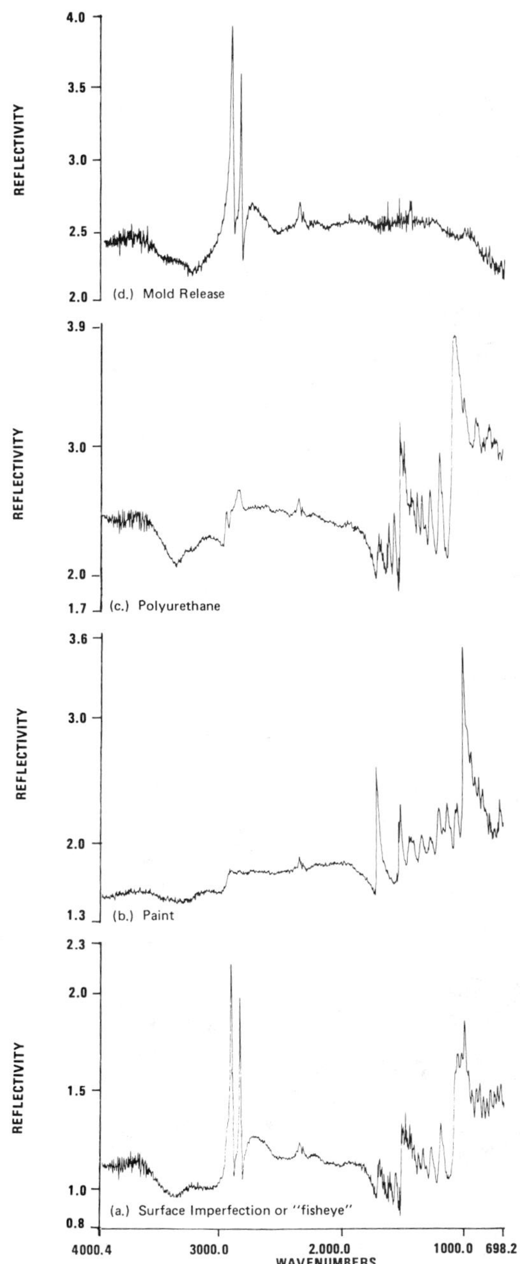

Figure 10. Infrared reflectance spectra representative of the (a) surface imperfections or "fisheyes", (b) paint, (c) polyurethane material, and (d) the mold release (reproduced from reference 2).

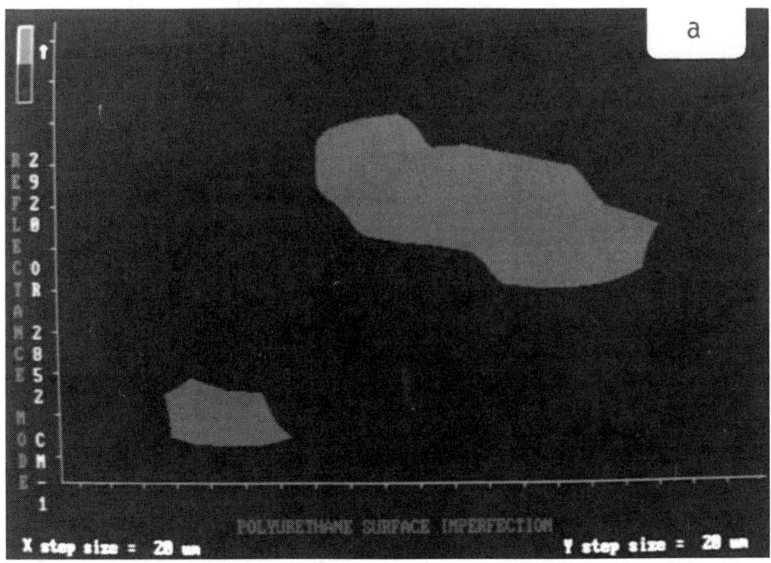

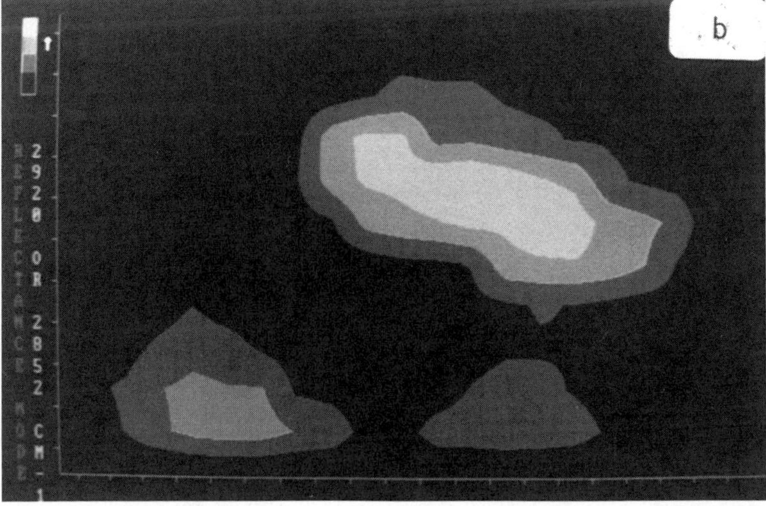

Figure 11. Functional group images of the surface imperfection in the polyurethane material shown in Figure 9. The images are based on 2920 or 2852 cm^{-1} absorptions and are represented by gray level images. The number of contour levels, L, as defined in Figure 2 are (a) 2, (b) 4, (c) 6, (d) 8, and (e) 30.

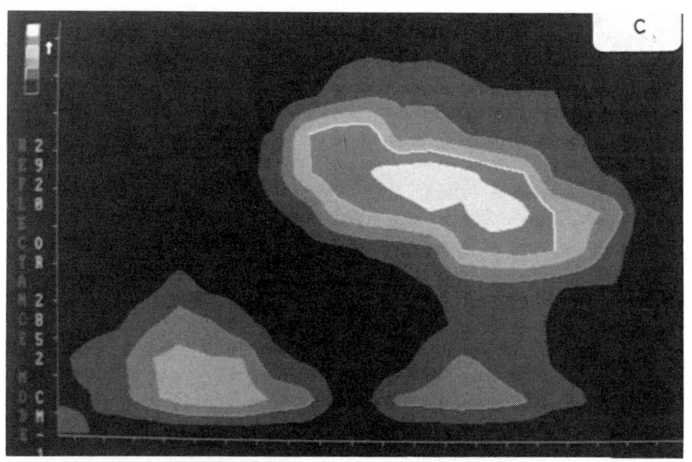

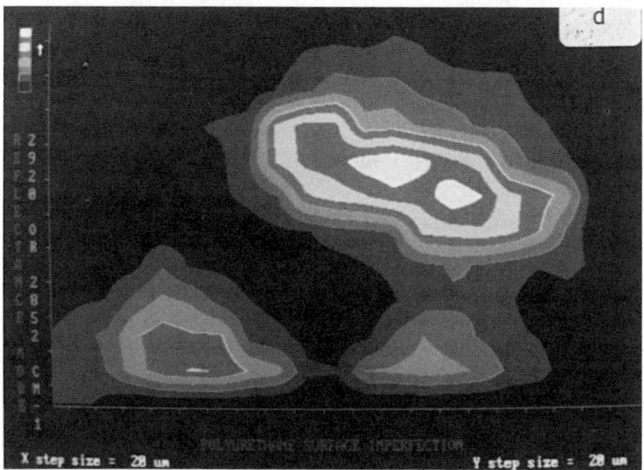

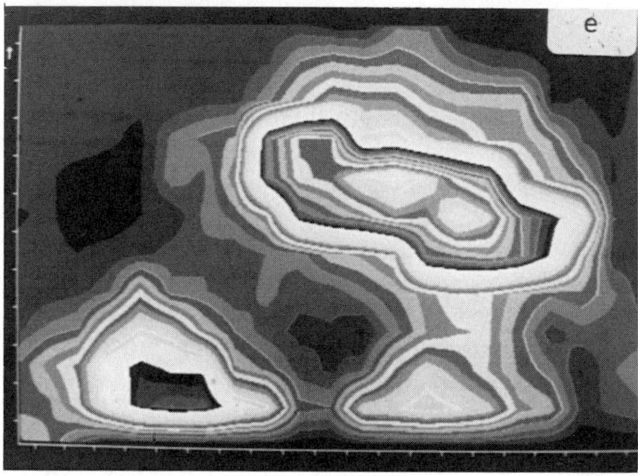

Figure 11. (continued)

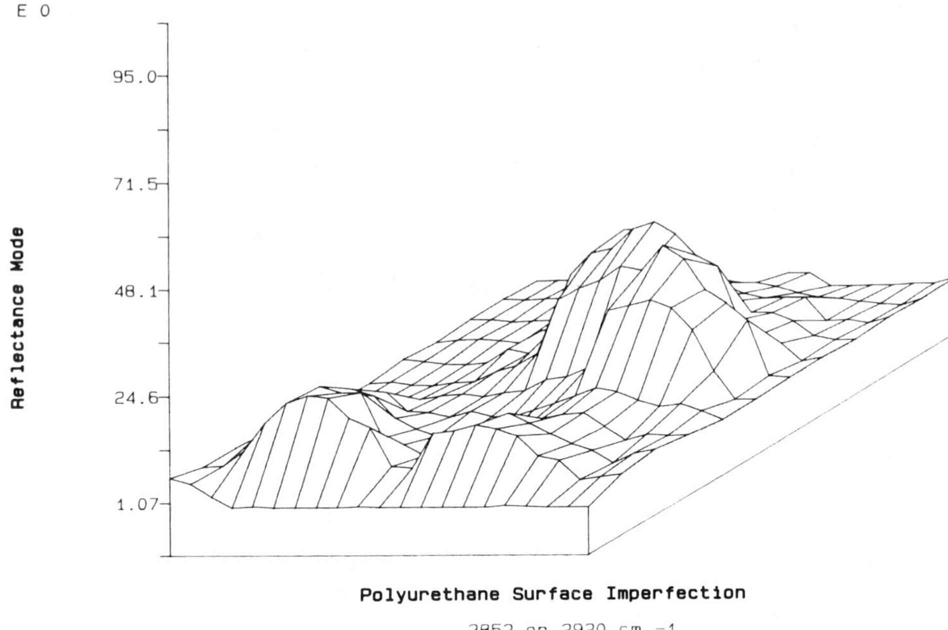

Figure 12. Functional group image of the surface imperfection in the polyurethane material shown in Figure 9. The image is based on 2920 or 2852 cm^{-1} absorptions and is represented by an axonometric plot.

2.4 CONCLUSIONS

The development and application of functional group imaging using infrared microspectroscopy has been discussed in this paper. The examples used here were to demonstrate several advantages and uses of the technique: automated sample collection, representative sampling of microscopic materials and imaging based on the use of functional group maps. The full impact and utilization of this technology has not yet been realized but many fields of science will benefit. For example, the technique can provide specific functional group maps, distributions, etc. whereas other techniques can only provide elemental information (e.g.electron microprobe).

The effective spatial resolution for which two different materials can be distinguished can be improved to 2-3 micrometers using the technique. The use of both experimental (see for example, reference 11) and mathematical [12] methods are being pursued to further improve the absolute and effective spatial resolution for which images can be obtained. Improved spatial resolution to below a micrometer can provide new discoveries in, for example, the biological, medical, and material science fields.

ACKNOWLEDGEMENTS

The authors wish to thank G. P. Young of the microscopy laboratory in the Analytical and Engineering Sciences Department of Dow Chemical, Texas Operations for preparing the microtomed cross-section of the multilayer polymer structure and the Dow Chemical Company for supporting the research presented in this paper.

REFERENCES

1. J. E. Katon, G. E. Pacey, and J. F. O'Keefe, *Analytical Chemistry*, 58: 465A-481A (1986).
2. M. A. Harthcock, L. A. Lentz, B. L. Davis, and K. Krishnan, *Appl. Spectrosc.*, 40: 210-214 (1986).
3. P. B. Roush, editor, *The Design,Sample Handling and Applications of Infrared Microscopes*, ASTM STP 949, American Society for Testing and Materials, Philadelphia (1987).
4. T. Hirschfeld, *Microbeam Analy.*, 246-252 (1982).
5. R. B. Marienenko, R. L. Myklebust, D. S. Bright, and D. E. Newbury, *Microbeam Analy.*, 159-162 (1985).
6. D. E. Newbury, *Microbeam Analy.*, 204-208 (1985).
7. S. L. Smith in *Instrumentation in Analytical Chemistry* (S. A. Borman, ed.), American Chemical Society, Washington, D. C. pp. 303-310 (1986).
8. R. P. Van Duyne, K. L. Haller, and R. I. Altkorn, *Chem. Phys. Lett.*, 126: 190-196 (1986).
9. T. J. Manuccia, *Paper #630*, Federation of Analytical Chemistry and Spectroscopy Societies Meeting XIII (1986).
10. R. G. Messerschmidt in *The Design, Sample Handling and Application of Infrared Microscopes* (P. B. Roush, ed.), ASTM STP 949, American Society for Testing and Materials, Philadelphia (1987).
11. G. A. Massey, J. A. Davis, S. M. Katnik, and E. Omon, *Appl. Opt.*, 24: 1498 (1985).
12. D. H. Burns, J. B. Callis, G. D. Christian, and E. R. Davidson, *Appl. Opt.*, 24: 154 (1985).

Analysis of Polymers by Infrared Microspectroscopy

3

The Identification and Characterization of Polymer Contaminants by Infrared Microspectroscopy

PATRICIA L. LANG*, J.E. KATON, ANTHONY S. BONANNO†, AND G.E. PACEY *Molecular Microspectroscopy Laboratory, Department of Chemistry, Miami University, Oxford, Ohio*

3.1 INTRODUCTION

A primary application of infrared microspectroscopy is its use in the identification and characterization of contaminants which occur in industrial processes or which contribute to material failure. Infrared microspectroscopy is well-suited for this type of analysis since selective, molecular information can be obtained on microscopic particles [1]. Using this technique, we have identified several particles removed from samples brought into our laboratory for contamination analysis. Most of the contaminants have been identified as polymers. More interestingly, many of these polymers appear to originate from settled dust and are not responsible for product rejection or failure.

In an attempt to characterize polymer contaminants with regard to their sources, the infrared spectroscopic study of particles in settled office and laboratory dust has been undertaken. Although dust contamination could occur in a variety of environments, an office or a lab seemed to be a common setting that all the materials might have experienced prior to being brought in for analysis. Preliminary findings suggest that polymers are indeed the major components of dust in this type of environment.

* *Current affiliation: Ball State University, Munsee, IN*
† *Current affiliation: Clarkson University, Potsdam, NY*

The purpose of this paper is twofold: (1) To make the reader aware of the presence, identification, and possible source of the many ubiquitous polymers which may interfere with the microspectroscopic analysis of materials, and (2) To demonstrate how such polymers can be distinguished from contaminants which are more likely responsible for material rejection or product failure.

3.2 EXPERIMENTAL

An Analect (Irvine, CA) AQS-20M system, consisting of a fXA-6260 spectrometer and a fXA-515 microscope module, was used to obtain the infrared spectra reported herein. Sample handling involved removing a particle from the contaminated material with a fine pointed probe and placing it on a 6-mm diameter KCl window. Particles approximately 10 μm thick or greater, and all fibers, were thinned by pressing with a probe on a microscope slide to improve spectral quality. Materials brought in for analysis were either examined immediately or stored in clean, plastic bags until analysis was completed.

Several different methods of dust collection were employed. One method involved pressing a glass slide on a table or shelf top. Dust particles which adhered to the slide were analysed. Another method involved scraping the dust from a surface with the edge of a glass slide. Dust was also collected by simply allowing it to settle on a glass slide. The latter method seemed better suited for the collection of smaller dust particles, while the former methods collected larger particles and fibers more readily. All slides were cleaned prior to use by spraying with pressurized dust remover. Dust was collected from several different offices and laboratories.

Spectra were not obtained on dust particles less than 10 μm. All spectra shown were untreated and consisted of 300 signal-averaged scans unless otherwise noted in the figure caption.

3.3 RESULTS

3.3.1 Dust Analysis-Polymers

Cellulose was one of the most prevalent constituents identified in the dust samples collected. Fibers from both wood and cotton were detected, as well as *particles* of wood cellulose. Figure 1 shows the spectrum of a wood cellulose fiber found in the settled dust. Wood cellulose can be distinguished from cotton cellulose by the presence of lignin. Characteristic lignin absorptions are

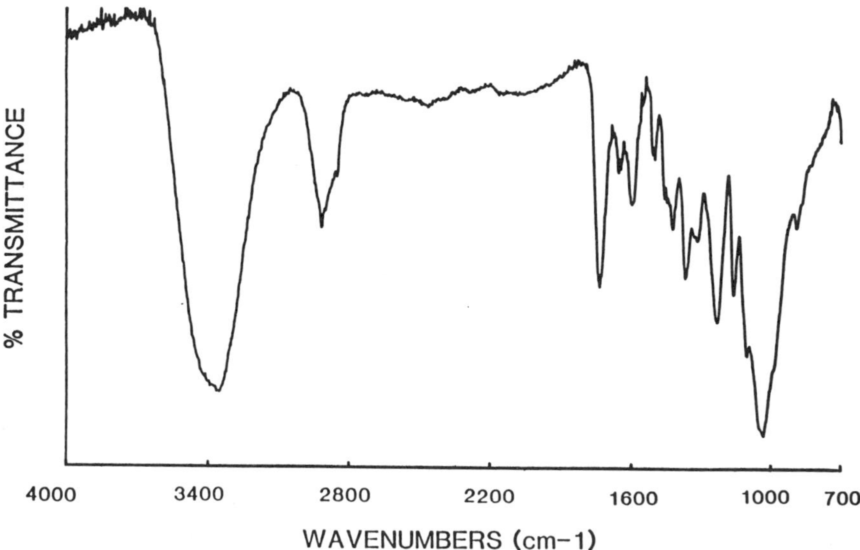

Figure 1. Spectrum of a cellulose fiber containing lignin found in settled dust.

observed at 1739, 1654, 1600, 1505, and 1235 cm^{-1} [2]. A likely source of wood cellulose as a dust particle is paper products manufactured from unbleached paper (e. g. corrugated boxes or newspaper).

Some of the cellulose identfied was found to contain kaolin. Kaolin is often used as a filler in the manufacture of paper products, and its presence suggests the source of these fibers or particles as paper. Kaolin can be easily identified in the infrared spectrum of a cellulose particle shown in Figure 2 by its characteristic absorptions at 3697 cm^{-1} and 3620 cm^{-1}. Kaolin also exhibits lower frequency absorptions; however, these bands are not easily observed due to the presence of cellulose. The absence of lignin bands in the spectrum of Figure 2 indicates a bleached paper as the source of this particular particle.

In the spectra of some cellulose fibers, only bands due to cellulose were present. Since the infrared spectrum of relatively pure cotton cellulose is undifferentiable from that obtained from well-bleached wood cellulose, microscopic appearance was used to identify these fibers. Characteristic convolutions which are present in unmercerized cotton fibers easily distinguish them from non-convoluted mercerized cotton or wood cellulose [3]. When convolutions were absent, cellulose fibers were not assigned to a specific source.

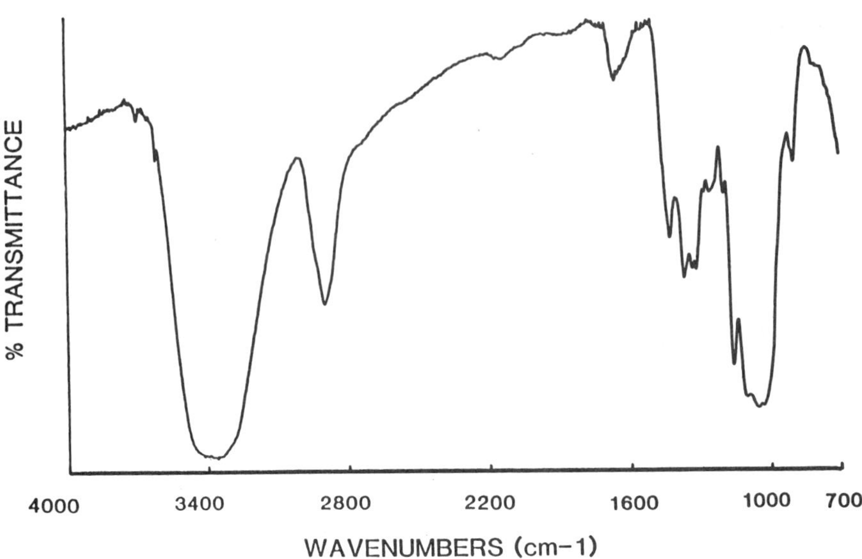

Figure 2. Spectrum of a dust particle: cellulose containing kaolin.

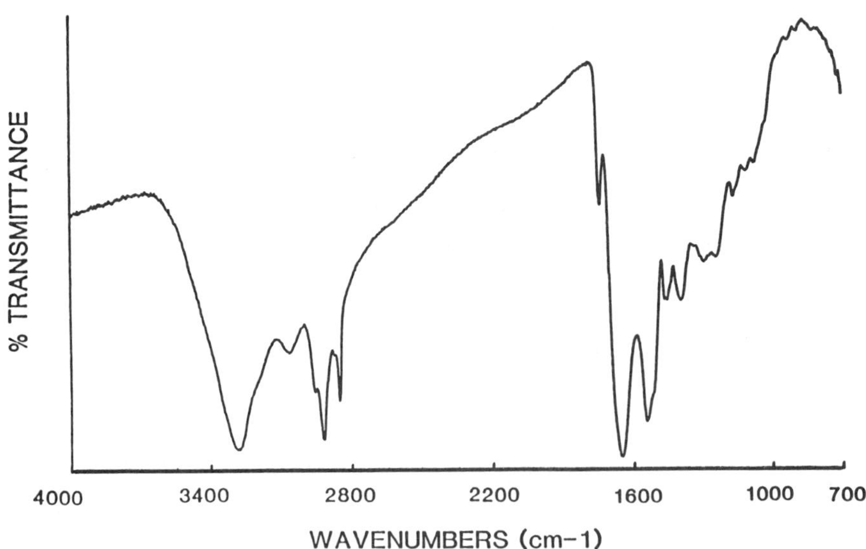

Figure 3. Spectrum of a dust particle: epithelial cell. (150 scans).

Polyamides were another prevalent component of the dust studied. The overwhelming majority of polyamides were present as flat, flaky particles. A representative spectrum of these particles, shown in Figure 3, matched spectra obtained from human skin scrapings. Under (400X) microscopic examination, these dust particles were recognized as epithelial cells [3], thereby confirming their skin origin.

Both natural and synthetic polyamide *fibers* were occasionally observed, too. They obviously originate from human or animal hair and synthetic textiles, respectively. As Figure 4 shows, the spectra of natural proteinaceous fibers differ from those of common, synthetic polyamides (e. g. nylons) not only by their peak locations, but by their general lack of spectral information in the lower frequency regions. In addition, natural proteins generally have a broader N-H stretching region and a less intense band at 3066 cm^{-1}, an overtone of the N-H bend. In addition, because of the more complex structure and composition of proteins in general, the C-H stretching region will often consist of three or four distinct bands. These characteristics can also be observed by comparing the nylon spectrum with the epithelial cell spectrum in Figure 3.

Other less frequently found polymers include polyethylene, polymethacrylates, polyvinyl chlorides, and binders typically used in paper or plastics manufacturing. The spectrum of one type of binder found is shown in

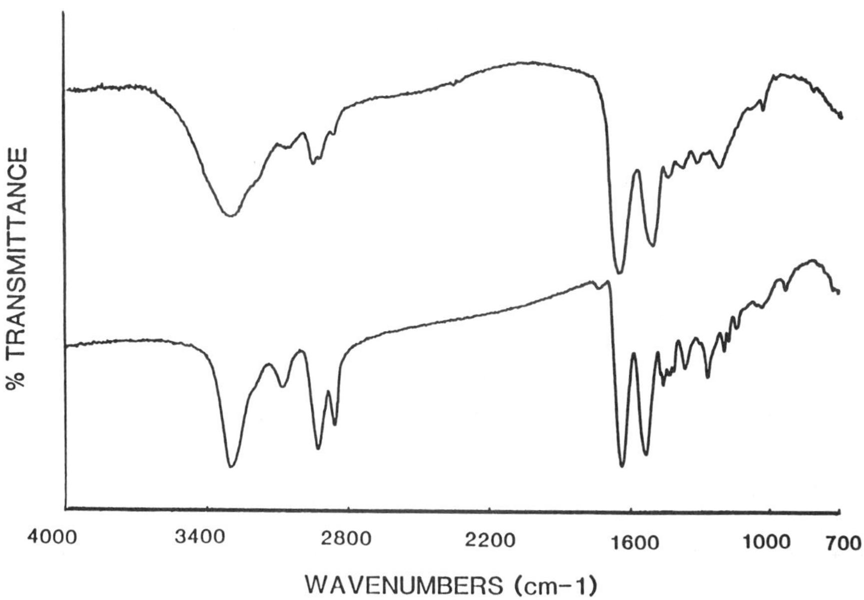

Figure 4. Spectrum of a hair fiber (top-131 scans) and a nylon fiber (bottom-159 scans), both found in settled dust.

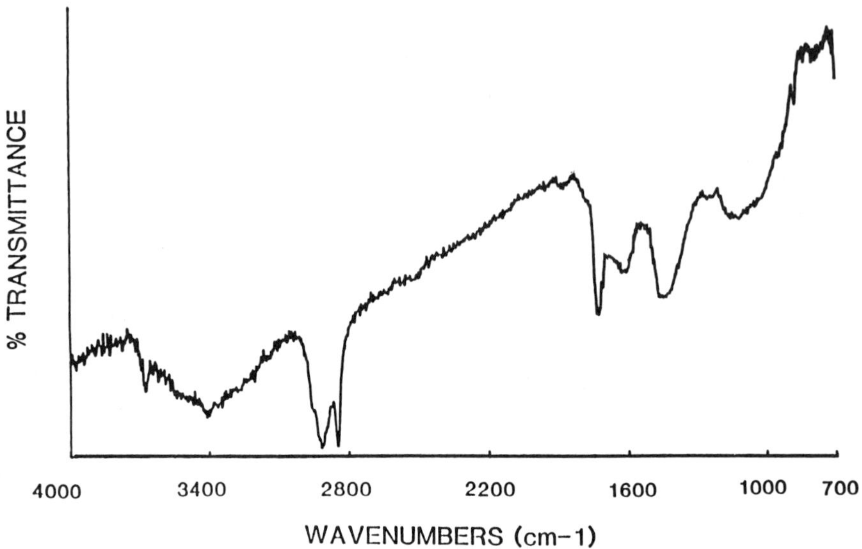

Figure 5. Spectrum of a brown dust particle identified as a mixture of components used typically as a binder.

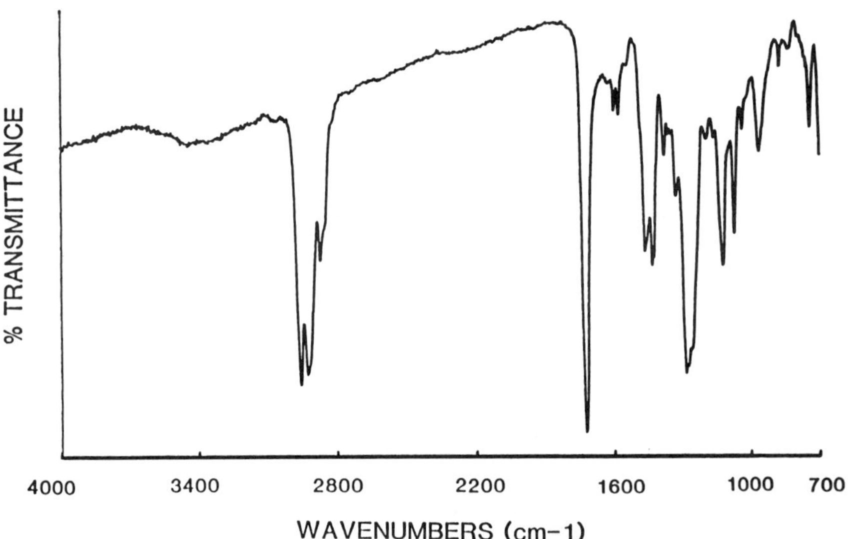

Figure 6. Spectrum of a dust particle: polyvinyl chloride plasticized with a phthalate ester.

Figure 5. Components of the binder which can be identified in the spectrum include clay, calcium carbonate, and a polyester, probably a polymethacrylate. Isolated polymethacrylates found in the dust samples may originate from the same sources. The polyvinyl chlorides detected contained a phthalate ester, compounds widely used as plasticizers. The spectrum of such a particle is shown in Figure 6. Different types of synthetic fibers such as polyester, glass, polypropylene, etc., could also be identified. Polyester fibers, originating from textiles, are the most frequently found fiber after cellulose. The glass and polypropylene fibers were only found in lab settings. Glass wool would be the likely source of the glass fiber. However, the polypropylene, present as a fiber, probably originated from a textile.

3.3.2 Material Analysis-Contaminants

Examples are given in the following paragraphs of analyses of different types of material for contaminants which might have caused material failure or product defect.

The first case involved small pieces of circuit boards which were to be analysed for the presence of solder flux residues. These pieces were suspect areas which had been cut from larger boards by the supplier after electronic failure had occured. However, upon microscopic examination, no residue was observed around the solder pads where flux would be expected to remain. Since the supplier of these boards suspected a water soluble flux, the extraction of any non-solid water soluble flux residues, such as polyglycols, was attempted [4]. The main constituents of the extracted residues were particles of cellulose, often containing lignin. Cellulose fibers and cellulose fibers containing kaolin were less frequently found. Naturally occuring polyamide particles were also present; their spectra matched those of epitheliel cell spectra discussed earlier. Although the boards were received and stored in plastic bags, the presence of this particular combination of polymers suggest that the extractions removed mostly dust particles which had settled on the board. Furthermore, the fact that no residues were present in localized areas on the boards and that celluloses and polyamides are not typically used in solder flux or circuit board manufacture support this assumption. The source of the unusually large amount of cellulose containing lignin particles present is unknown.

Other polymer particulates identified from the extraction included a polyfluorochlorocarbon, a paper binder, two types of polyphenoxy resins, and a polymethacrylate. The binder and one type of phenoxy resin appear to be components of the fiberglass substrate. The solder mask was the source of the other type of polyphenoxy resin found, while the polymethacrylate was used as a protective overcoating material. The polyfluorochlorocarbon may be residual solvent left from the board cleaning process.

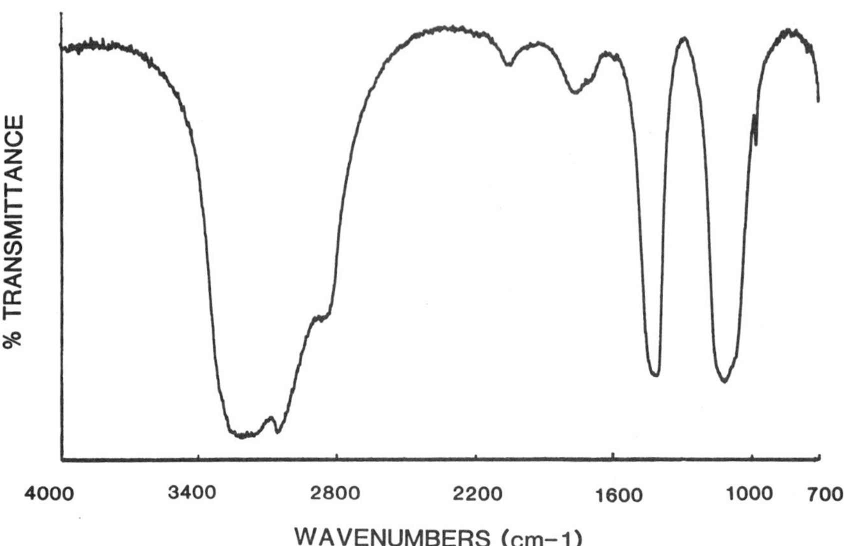

Figure 7. Spectrum of a metal activator containing ammonium chloride removed from a failed circuit board.

At a later date, larger sections of the boards were received. On one of the boards brown residue was apparent along a row of solder pads. The infrared spectrum shown in Figure 7 closely matched that of a metal activator containing ammonium chloride, a likely culprit responsible for the demise of the copper circuitry.

In the next example, sheets of window glass were analyzed because streaks of residue on the glass surface caused customer dissatisfaction. After scraping some of the residue off and obtaining its infrared spectrum, it could be identified as a naturally occuring polyamide. However, the residue did not consist of relatively large, flaky particles like those associated with skin. Instead, it was a thin, fairly uniform, film-like residue. The presence of *streaked* residues suggests that they were systematically formed during an assembly line process.

One possible source of these polyamides may be water baths used in the glass washing/rinsing process which had become contaminated with bacteria. Detergents containing protein additives may be another possible source. A slightly different type of window glass received also had streaked residues on its surface. These residues differed in appearance from those mentioned above in that they were much thinner, and sample quantity was extremely limited. A spectrum obtained from particles removed from the streaked area is shown in

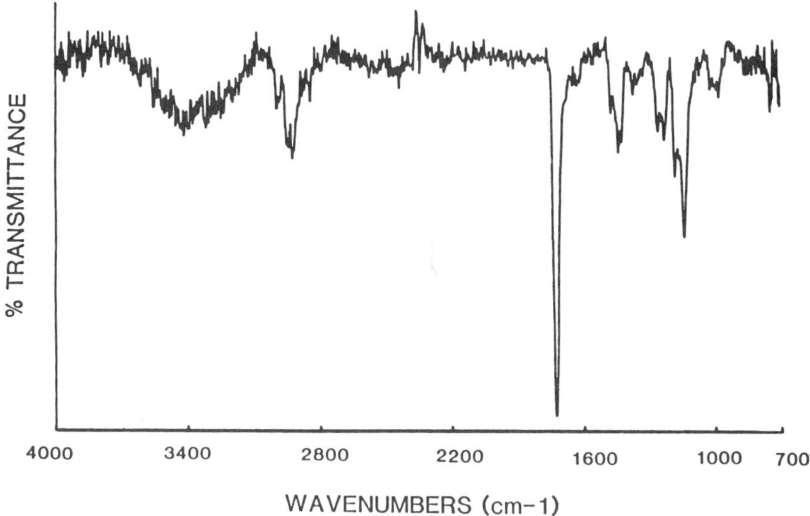

Figure 8. Spectrum of a polymethylmethacrylate particle approximately 10 µm in diameter removed from defective glass. (Baseline corrected-250 scans).

Figure 8. The spectrum can be identified as that belonging to a polymethylmethacrylate (PMMA). Since the manufacturer used PMMA beads as a packing material, the likely source of the residue was easily traced.

3.4 CONCLUSIONS

It is quite common to find dust contamination on materials brought in for microspectroscopic analysis, and usually such contamination is not the cause of a material defect or product failure. However, the presence of dust particles may, at best, temporarily obscure more meaningful data and, at worst, steer the analyst completely away from the solution to a problem of high priority.

Even if one is aware of the possible presence of such extraneous particles, precautions should be taken to avoid dust contamination.

Suppliers of samples to be analyzed should be cautioned to handle materials with tweezers or tongs, if possible, and to protect them from exposure to the environment. The same precautions should be taken in the microspectroscopy laboratory as well. If dust contamination is unavoidable, the analyst should look for contaminants which are arranged systematically on the material. If it is necessary, a detailed description of the material's history may be helpful in interpreting the data.

Although the present discussion is primarily restricted to polymers, it is not meant to imply that non-polymeric particles play insignificant roles in industrial process or as constituents of dust. Sulfates, carbonates, and silicates, for example, are significant in both respects. These particles will be included in a more comprehensive paper, along with Raman microspectroscopic data, at a future date [5].

REFERENCES

1. J. E. Katon, G. E. Pacey, and J. F. O'Keefe, *Anal. Chem.,* 58: 465A-481A (1986).
2. D. O. Hummel, and F. Scholl, *Atlas of Polymer and Plastics Analysis*, Vol. 2, Part a/I, Carl Harser Verlag, Munich (1984).
3. W. C. McCrone, R. G. Draftz, and J. G. Delly, *The Particle Atlas*, Ann Arbor Science Publishers, Ann Arbor, MI (1967).
4. P. L. Lang, and J. E. Katon, *Microbeam Analysis*, 47-49 (1986).
5. P. L. Lang, J. E. Katon, and A. S. Bonanno, *Appl. Spectrosc.*, accepted for publication.

4

Polymers and Contaminants by Infrared Microspectroscopy

HOWARD J. HUMECKI *Walter C. McCrone Associates, Inc., Chicago, Illinois*

4.1 INTRODUCTION

Infrared spectroscopy has made giant strides in the last ten years and nowhere is it more in evidence than in the analysis of small samples. Just six years ago, we were using beam condensers with pinhole apertures to examine 100 to 200 micrometer sized specimens. Now we can routinely analyze particles of 25 micrometers in size. A few years ago, sample size was generally accepted to be diffraction limited. We now realize that while imaging may be diffraction limited, we can effectively reduce this limit for analytical purposes. Expectations are that we will be able to prepare spectra of sub five-micrometer specimens and perhaps down to within one micrometer in a matter of five years or so. The extending of our range of capabilities has been largely the result of technological improvements, improvements in hardware, improvements in software and in accessory design. The beam condenser and pinhole aperture have given way to sophisticated microscopes with Cassegrainian lens systems and facilities for transmission and reflecting modes.

We can do so much more as a result of these developments that we tend to rely almost totally upon them and less and less upon our own skill and ingenuity. We all are in a rut and are reluctant to deviate from standard procedures. If our spectrum is poor, we tweak the instruments or we increase the number of scans. We use reflectance instead of transmission or *vice versa*.

After we have used our instruments for a period of time, we know pretty much what we can and can't do. The design that we are given is a constraint that we have to live with. We have to work within it and unless we redesign it, we aren't going to exceed those capabilities. That is why, with few exceptions, we generally know, before we start, if we are likely to be successful or not with a particular specimen. Once we have optimized our instruments, there is very little that we can do with them, but we can bring our specimens within the performance range of the instrument.

It seems that the last thing that we are inclined to do is touch the sample itself, and yet we must manipulate or modify it in order to bring it within the performance capabilities of the instrument in order to be successful. We prefer to work with a sample as we find it, or in a matrix. Maybe we lack the confidence in our manual dexterity. Whatever the reason, if we don't take advantage of the sample handling techniques available to us now, and develop new ones, we are going to find ourselves with technological overkill and still be unable to solve the more difficult problems that come along.

The people most skilled to handle and cope with small samples are microscopists. They deal with them everyday and are expert at knowing how to manipulate them. We have to learn to apply some of the techniques that they use and combine them with some generally accepted practices of the spectroscopists.

4.2 EXPERIMENTAL

The instrument used for the examples reported here is a Digilab (Cambridge, MA) Model FTS-20C FT-IR equipped with an old Perkin-Elmer (Ridgefield, CT) Model 85 microscope mounted on top (Figure 1). A mirror in the sample chamber diverts the beam through the large Cassegrainian condenser lens of the microscope. Although the microscope is 35 or 40 years old, we believe that it is comparable in performance to currently commercially available microscope attachments. The microscope does not have facilities for polarizing the light beam.

The tungsten needle is the bread and butter tool of the microscopist. It is shown in the lower right of Figure 2. Its point is about one micrometer in size. The upper image is that of a common sewing needle. One must have the proper tools for manipulating small particles and the sharp tungsten needle is that tool. It can be seen that if one tries to deal with some of the particles in the figure, they are going to get lost. They will stick to the needle tip and be difficult to remove or see. Sooner or later they are going to fly somewhere into space, never to be seen again.

Figure 1. Microscope attachment used in experiments.

Tungsten needles are made of a tungsten wire about 0.6 millimeters in diameter. The tip has been etched or sharpened in molten potassium nitrite or prepared electrolytically in a caustic bath. It takes very little time to prepare them and they are absolutely essential for the manipulation of small particles [1].

4.3 EXAMPLES

Figure 3 shows a microscope slide with three strips of adhesive tape across it. Each one has a streak in it. These were prepared by pressing the tape against the siding of a house in a Midwestern region and pulling the tape away

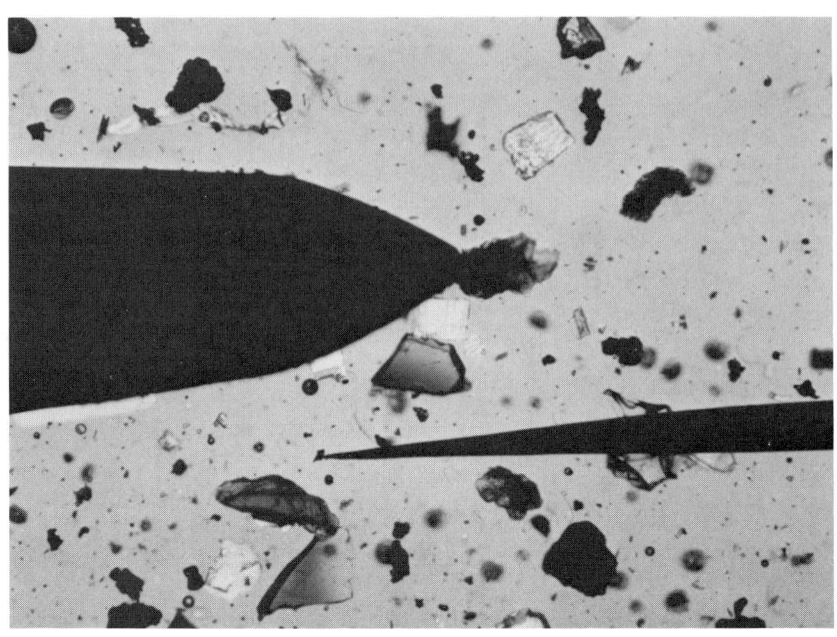

Figure 2. Tungsten needle having 1 μm tip compared to tip of sewing needle.

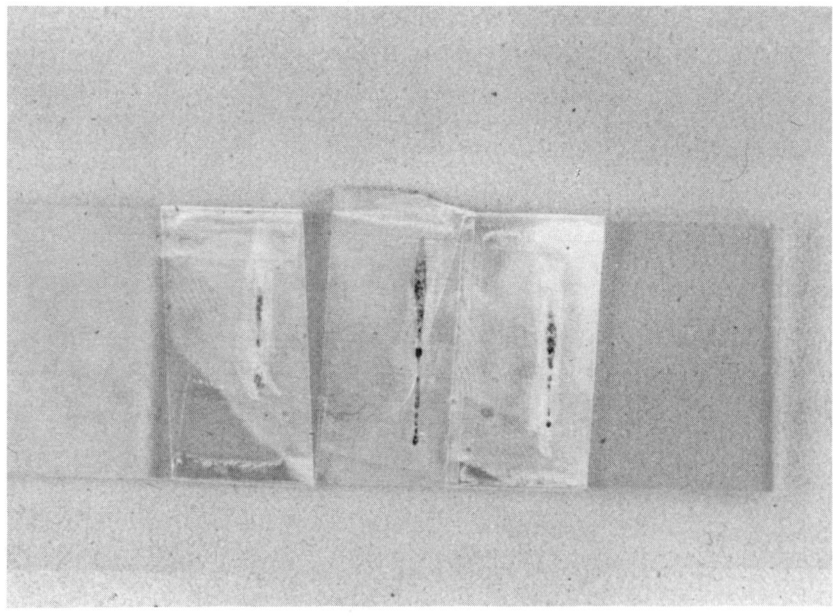

Figure 3. Microscope slide with three sticky-tape sample streaks.

with the streaks attached. The narrative involves a refinery which was blamed for spewing out a blackish viscous liquid that was falling onto local homes producing the streaks. Most people blamed the refinery, but some also noticed large numbers of spiders in the region. They attributed the streaks to deposits of the spiders.

When the samples were sent to us by the refinery, they were found to consist of a patch of brownish, partially crystalline particles with a blackish edge. Although the adhesive tape was effective in removing the streaks from the siding, it was very difficult to separate the deposit without mixing it with the adhesive from the tape. But with careful manipulation, using a pointed scalpel and a tungsten needle, we were able to remove enough material to obtain the spectrum shown in Figure 4 (top). This substance was identified as guanine, one of the major constituents of bird and bat guano as well as the excretory products of spiders. Figure 4 (bottom) is that of a reference grade guanine. You can see an almost perfect match with no extraneous absorption bands from the adhesive tape. The refinery was off the hook. The streaks were not the result of deposits from the refinery, but the excreta from spiders, or, perhaps, even the bats that eat the spiders and other insects.

Supporting evidence was obtained by elemental analysis of the deposit, employing the electron microprobe. Calcium, phosphorus, potassium and some other minor elements were detected. All are consistent with the excretory products of spiders, birds and bats. This illustration shows that even a sample made difficult by mixture with an unwanted matrix can be isolated in relatively pure form by careful manipulation.

Figure 5 represents a particularly difficult problem. These crystals developed in a three-layer laminate of plastic. They are only observable under polarized light. In addition, the crystals are feathery and very fragile. An attempt to remove them in a pure form, even by a skilled microscopist, was unsuccessful. We were able to remove some of the crystalline material with a minimum of the center layer. It was in the center layer that the crystals were observed. Although we were able to remove the crystal in the polymer matrix, this had to be done under slightly uncrossed polars in order to observe the crystals.

Since this infrared microscope attachment had no provisions for observing the sample under polarized light, we had to improvise. A plastic polarizer was taped to the entrance to the substage condenser lens of the microscope and the analyzer was taped to the entrance to the objective lens in slightly uncrossed position. In that way we could locate the crystals and

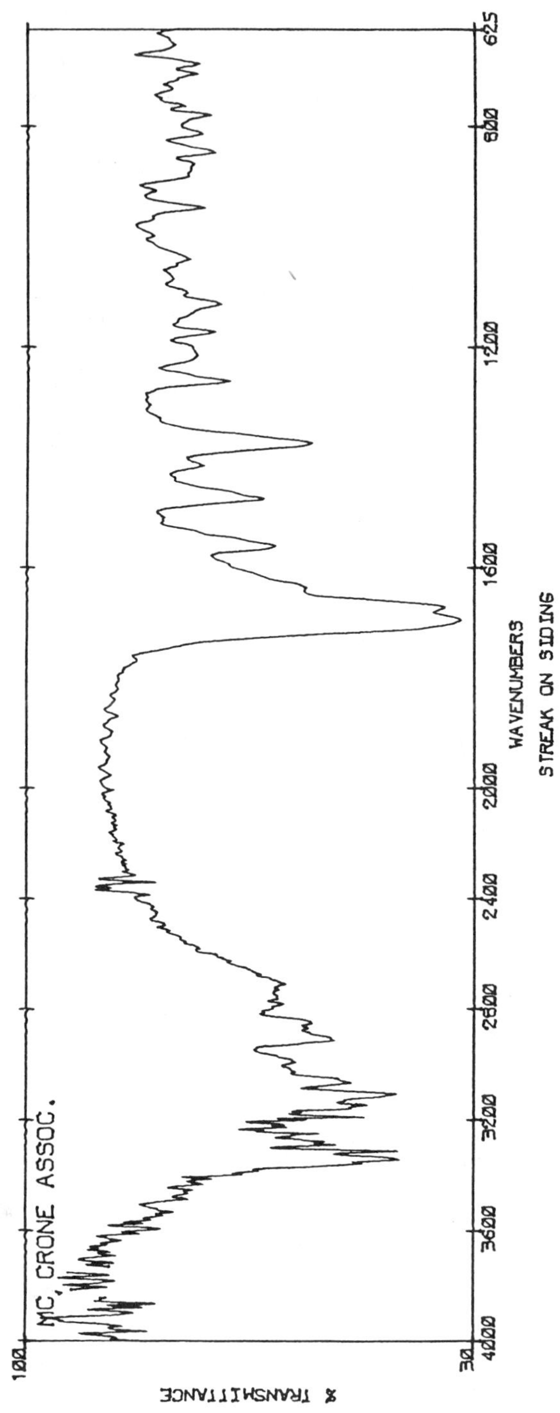
MC CRONE ASSOC.
% TRANSMITTANCE
100
30
4000
3600
3200
2800
2400
2000
1600
1200
800
625
WAVENUMBERS
STREAK ON SIDING

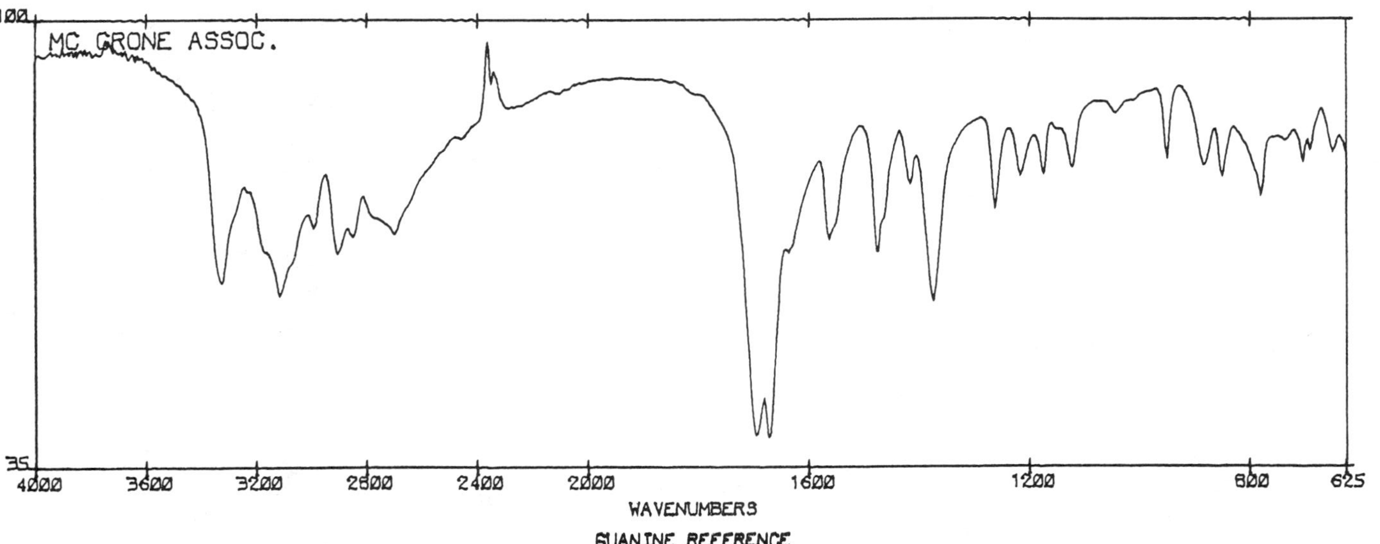

Figure 4. Siding residue (top). Reference guanine (bottom).

aperture them sufficiently to get a high concentration in the beam. After the aperture was determined, the polarizer and analyzer were removed and the specimen was analyzed. A spectrum of the matrix polymer was then prepared and subtracted from that of the crystal contaminated with matrix. Figure 6 shows the subtracted spectrum as well as that of a reference behenamide. Behenamide is one of the polymer additives which was apparently crystallizing from the matrix. This illustration shows that even if you cannot obtain an absolutely pure specimen, you can improve the spectral quality to the point where it can be identified.

More often than not, we deal with particles that are equant rather than flat, thin films. If these equant particles are analyzed without modification, we can lose energy by diffusion and scattering. Sometimes the samples will be too thick and the spectra will "bottom out". We can change the shape of these particles to improve spectral quality.

Figure 7 is an SEM micrograph of a particle approximately 15 x 30 micrometers in size, that was a contaminant in a molded part. Figure 8 is a spectrum of that particle, without modification. The spectrum is poor, with only the suggestion of a polyamide-type character. The specimen was removed and pressed between two polished steel plates at approximately 8,000 psi.

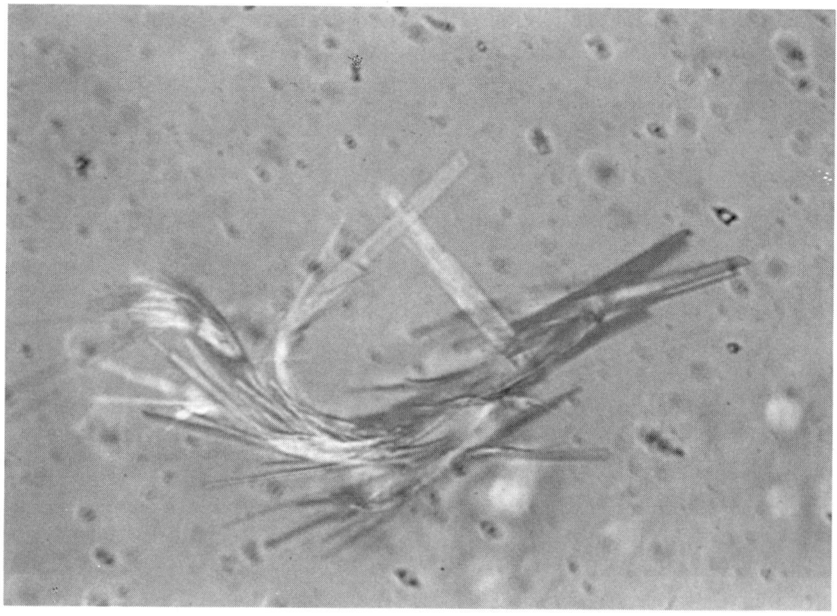

Figure 5. Crystals found in plastic film. Slightly uncrossed polars.

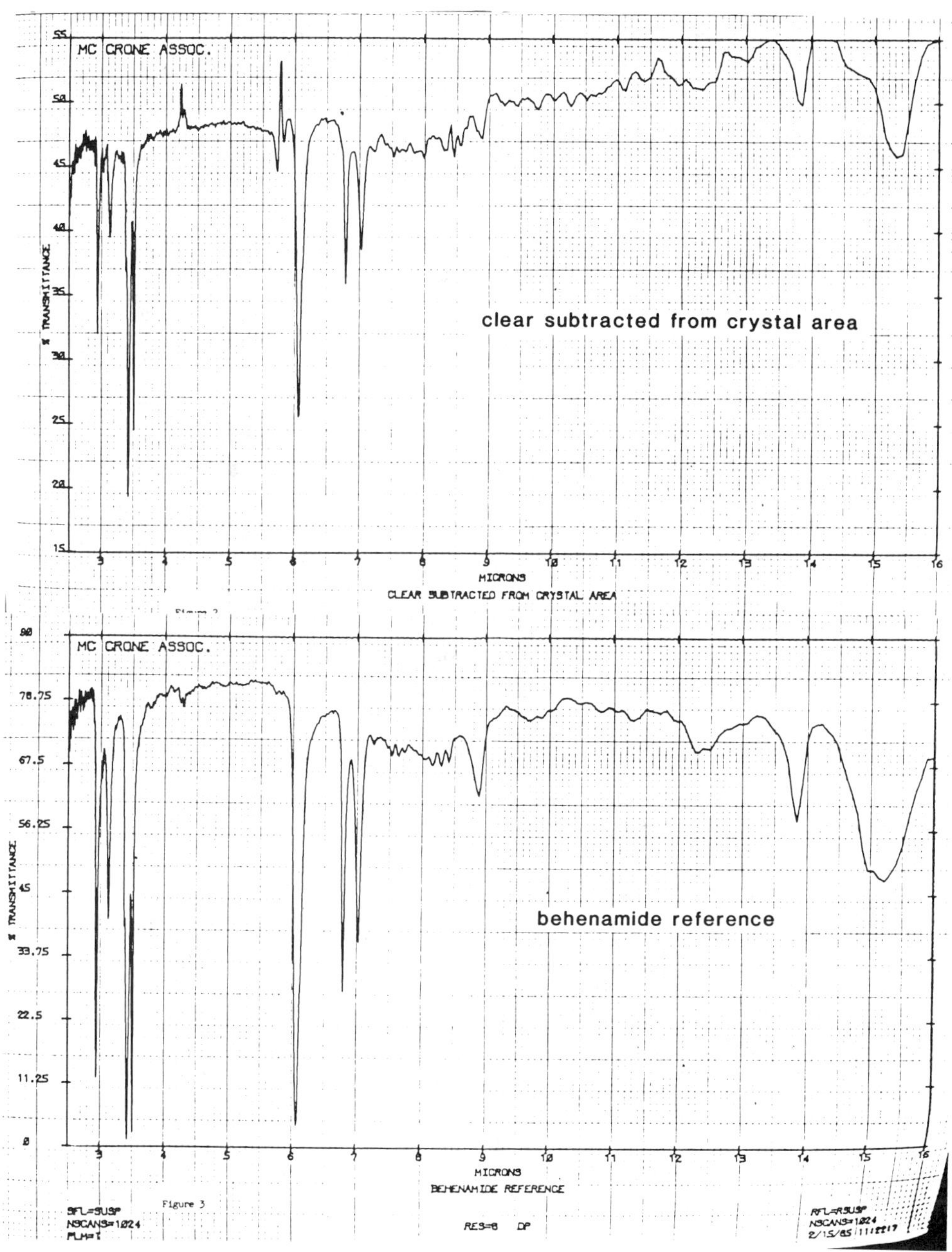

Figure 6. Matrix polymer subtracted from crystal region (top). Reference behenamide (bottom).

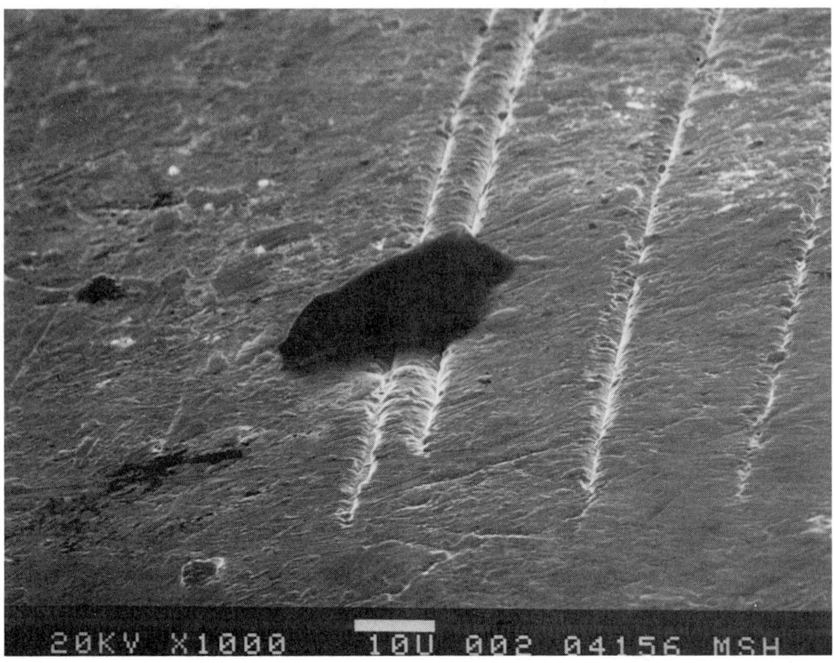

Figure 7. Particle found in molded part.

Although this technique is not very elegant, it is much cheaper than a diamond anvil and the results are very good. Figure 9 shows the particle after pressing. The particle is approximately 60 to 80 micrometers in diameter and, although fractured, the fracture did not interfere with the analysis and the spectrum shown in Figure 10 is considerably improved. The specimen is readily identified as a nylon. By the simple process of pressing the sample, we increase the cross section, diminish scatter and, consequently, increase the energy throughput.

Figure 11 is a thin section of heart tissue removed from a patient at autopsy. Under slightly uncrossed polars, birefringent and non-birefringent disks can be observed. They represent cross sections of surgical thread. Tungsten needles were used to remove some of these disks, which were approximately 20 micrometers in diameter. Although they could have been analyzed directly, we chose to press them between steel plates and obtained the

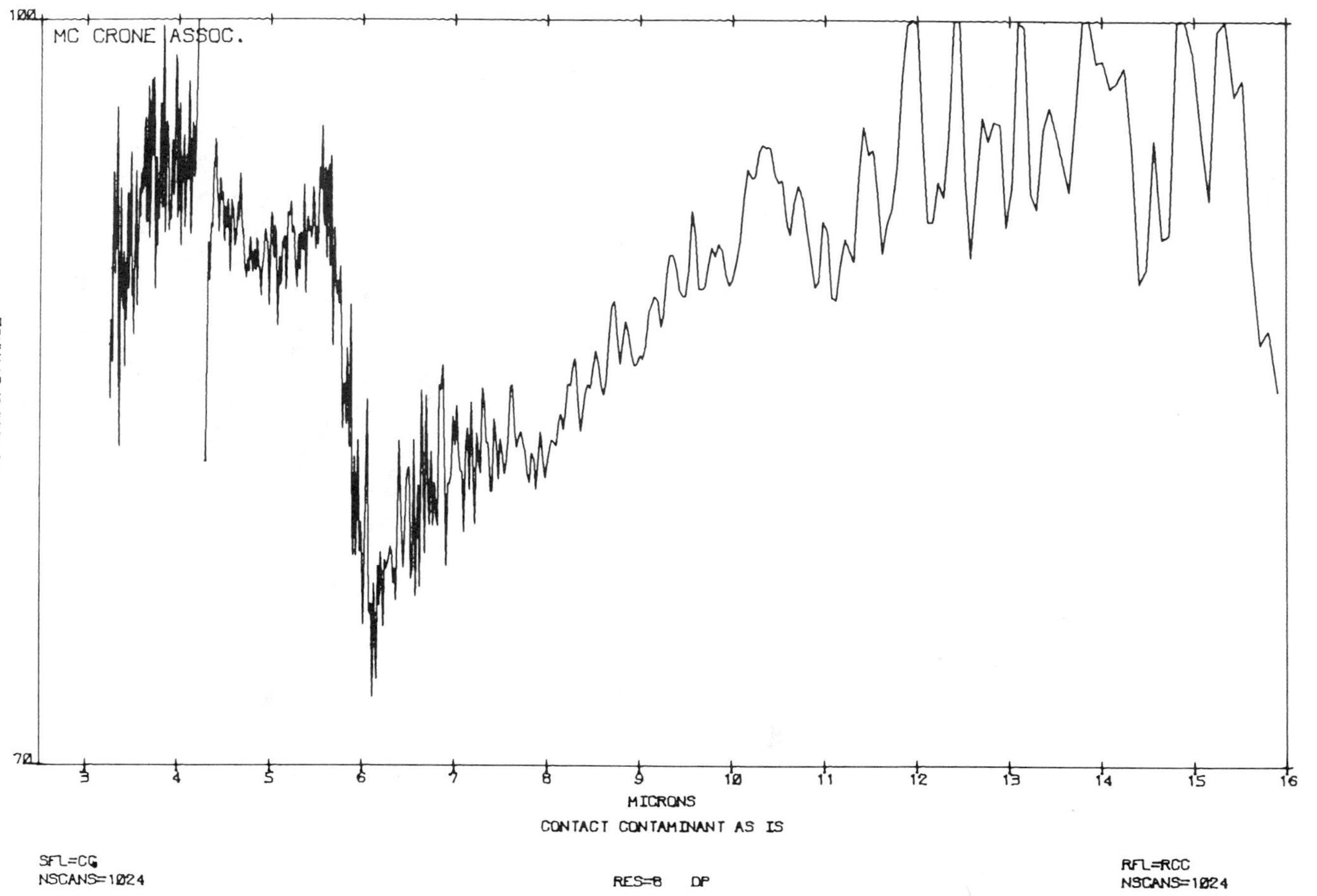

Figure 8. Spectrum of particle shown in Figure 7.

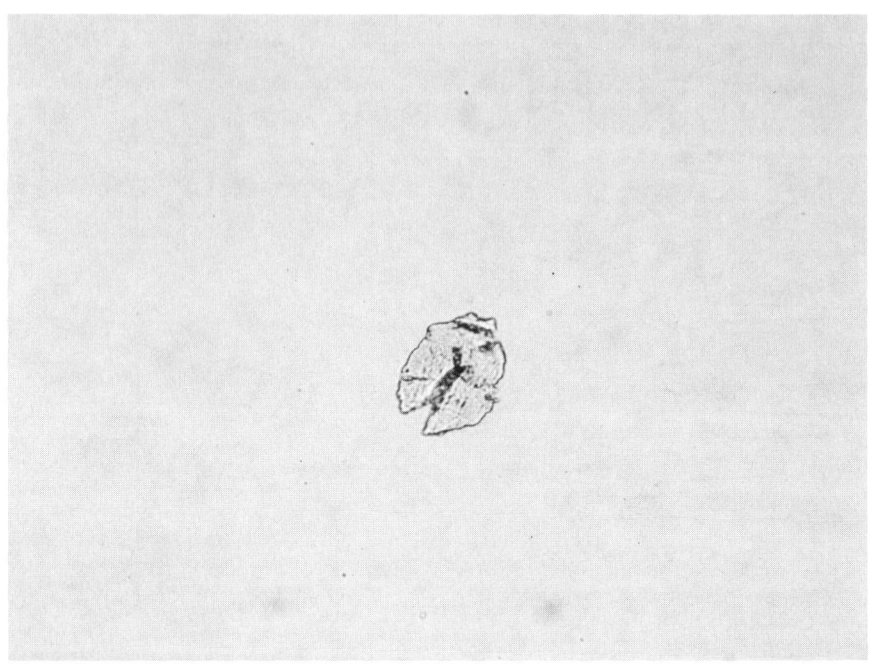

Figure 9. Photomicrograph of particle after pressing between steel plates.

spectra shown in Figures 12 and 13. They represent polypropylene and Teflon®-type polymers. Incidentally, sometimes we will examine "clean" disks and find on examination, a large number of dust or fiber particles on their surfaces. These can be easily removed by pressing adhesive tape to the surface and then removing the tape and its attached particles. When the specimen is exceptionally small and we are concerned about losing it on the 1" diameter of the steel disk, we will draw a circle on the disk enclosing the particle so that we can relocate it. If you separate the two disks as you would book pages, you will have no difficulty in finding the particle, either within the circle or on the opposite disk in the reverse image.

Powdery samples scatter a lot of energy and if there is insufficient sample for a small KBr pellet, we do what microscopists have been doing for well over a hundred years and that is to use an immersion oil. Mineral oil is ideal because of its few absorption bands. Spectrum 14 (top) is that of an anti-oxidant scraped from a surface and deposited directly on a salt plate. The background

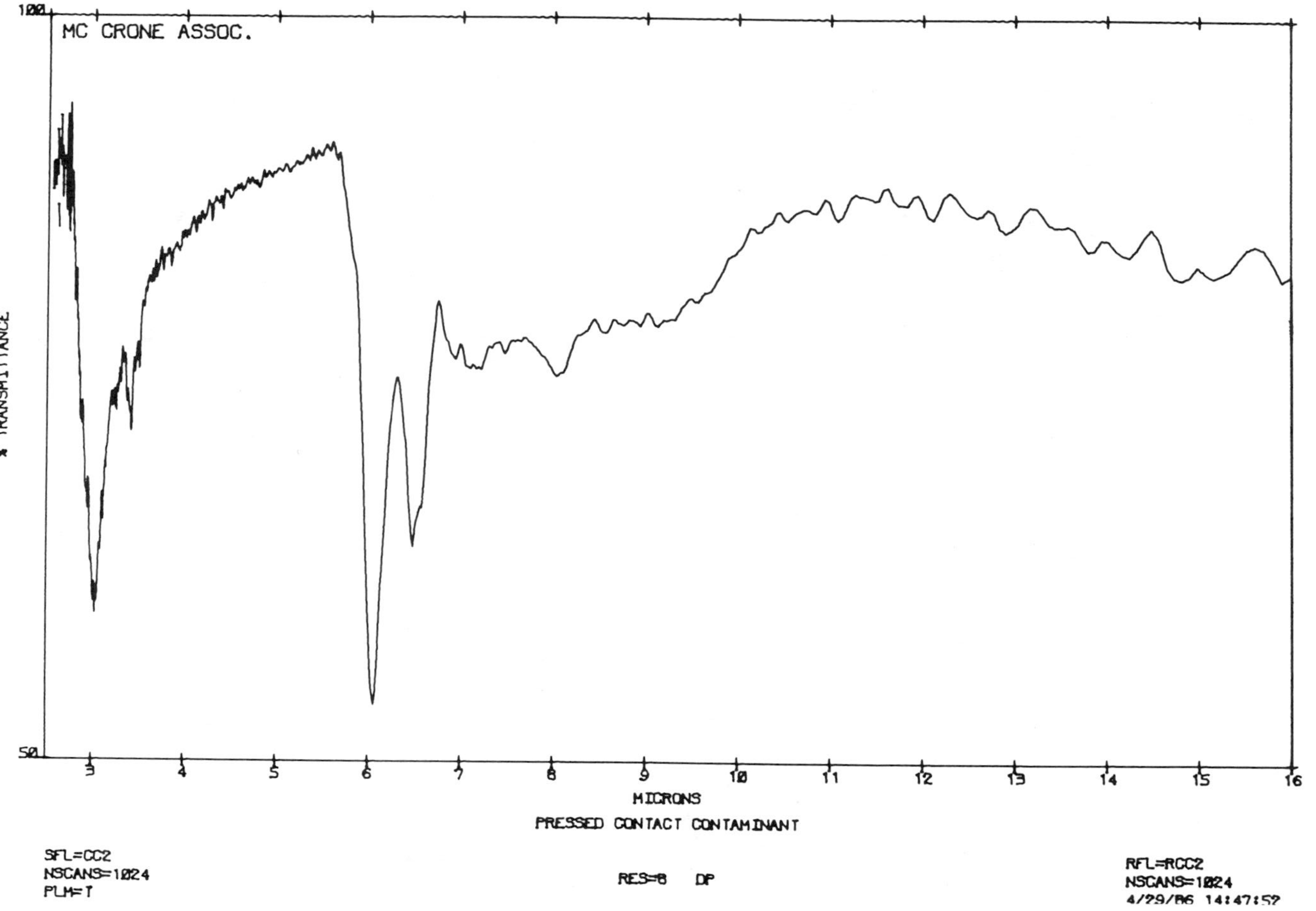

Figure 10. Spectrum of particle, after pressing.

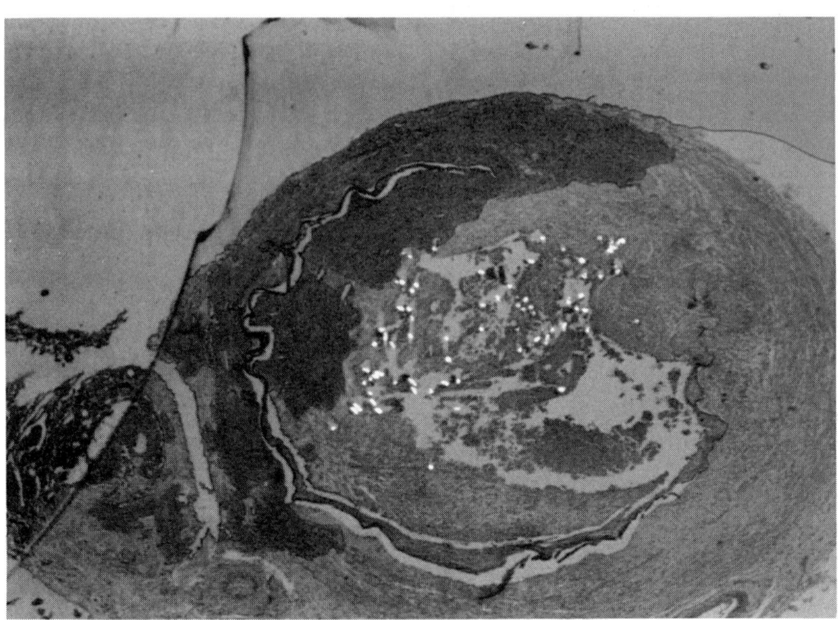

Figure 11. Thin section of heart tissue. Slightly uncrossed polars.

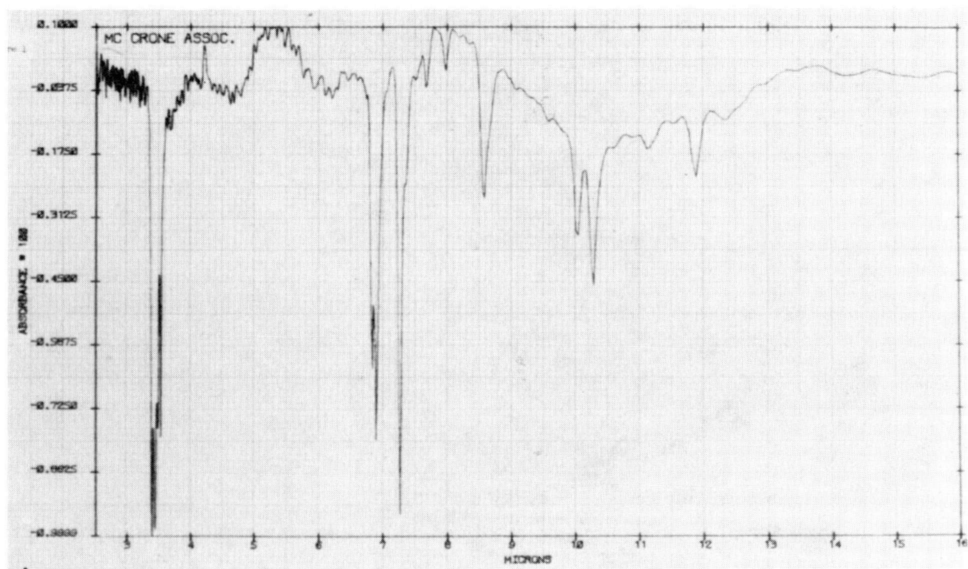

Figure 12. Spectrum of pressed surgical thread cross section (polypropylene).

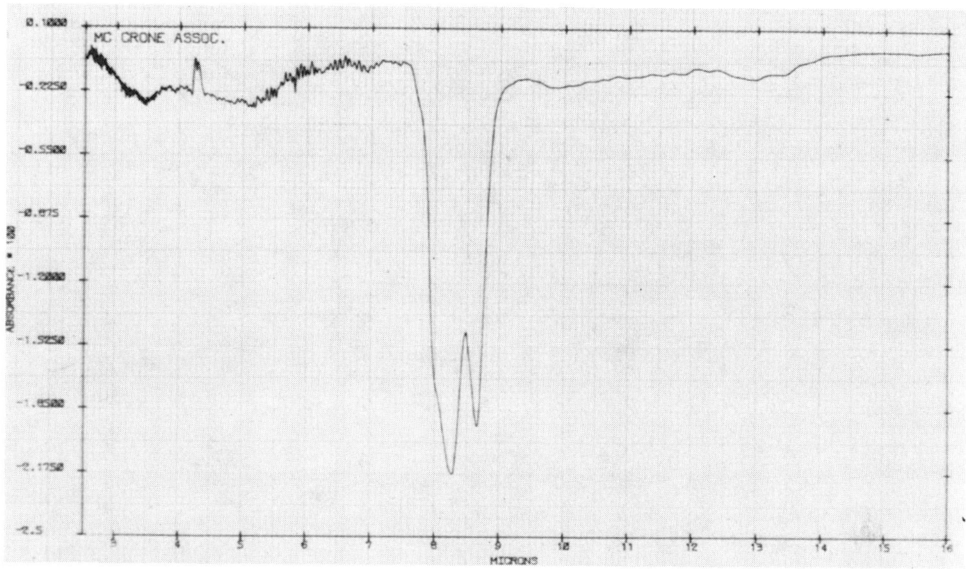

Figure 13. Spectrum of pressed surgical thread cross section (Teflon®-type).

was then subtracted from it. Although the spectrum is sufficient for identification purposes, it leaves something to be desired. By placing a small salt disk over the sample and then applying a drop of mineral oil between salt plates, the spectral quality can be improved. Spectrum 14 (bottom) shows that of the immersed specimen with the mineral oil spectrum subtracted from it. Again, the technique is very simple, but it does result in an improvement. In addition, it works best with the most difficult samples, i.e., those that are dark in color and have the highest refractive index or those that scatter or absorb the most light.

The salt plates, upon which the samples are mounted, are prepared from large sodium chloride or potassium bromide crystals that have been cleaved to one-half to one millimeter thickness. They are smoothed with a #600 Emery paper and then polished on a paper towel that has been wet with a few drops of alcohol/water. Surfaces can be repolished by gently wiping them with a pipe cleaner or Q-tip® that has been wet with an alcohol/water mixture.

One of the most frustrating problems that we run into involves little globs of elastomers or adhesive that are too thick to analyze directly and yet too soft to slice with a scalpel. When they are pressed onto a salt plate, they keep popping up again, without becoming any thinner. We have solved this problem

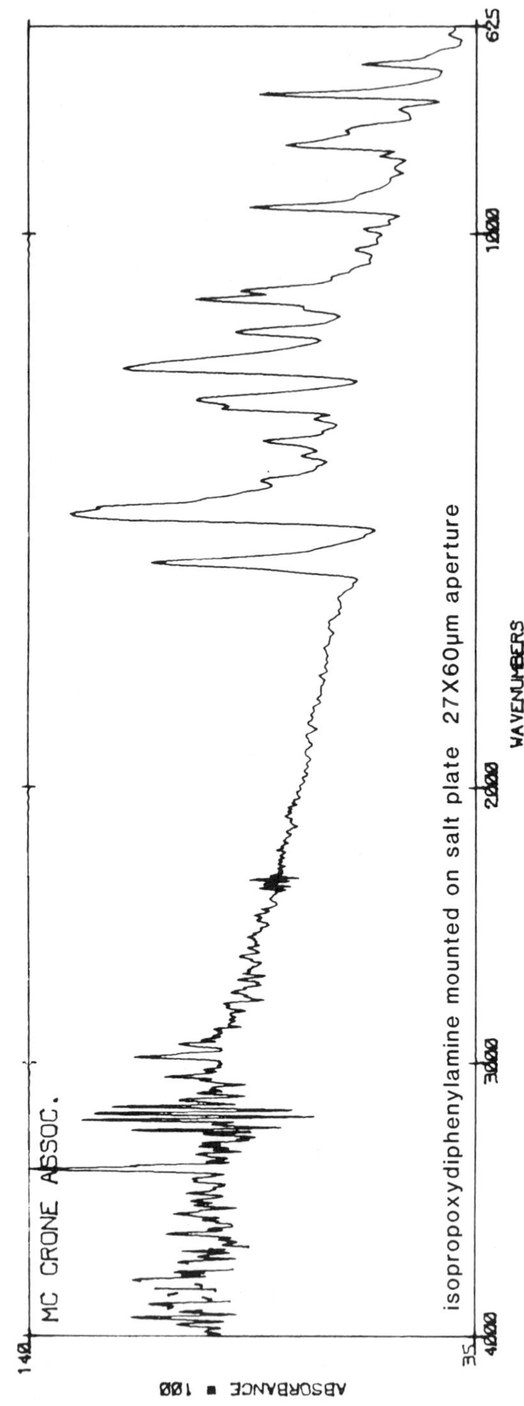
MC CRONE ASSOC.
isopropoxydiphenylamine mounted on salt plate 27X60µm aperture
WAVENUMBERS
ABSORBANCE * 100
140
35
4000
3000
2000
1000
625

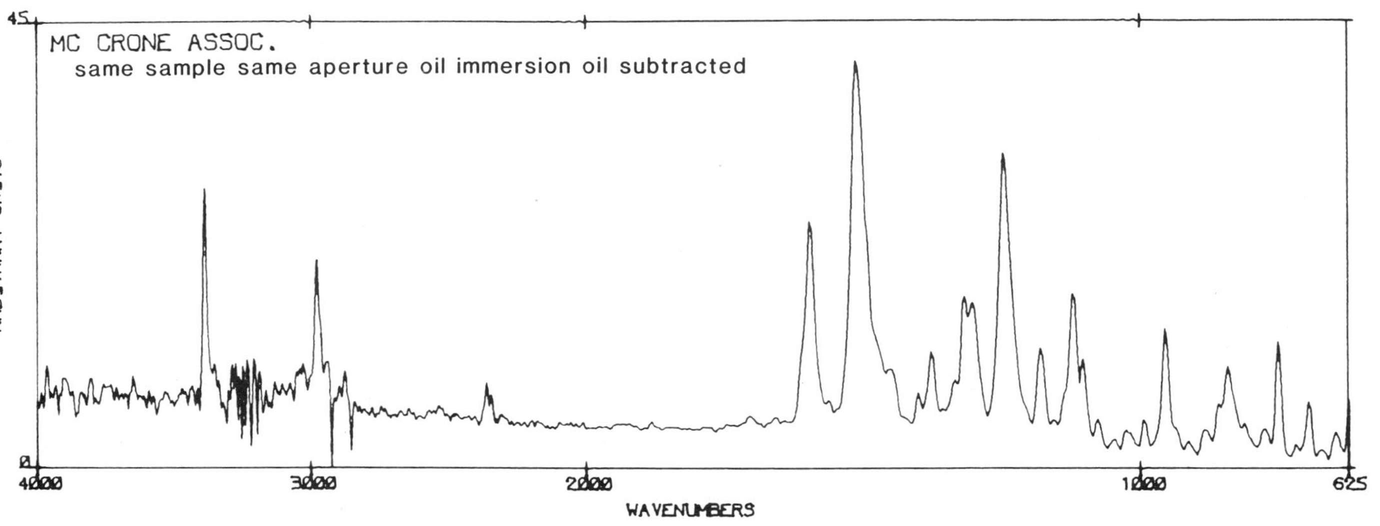

Figure 14. Spectrum of anti-oxidant (top). Spectrum of same specimen in mineral oil after subtraction of the oil spectrum (bottom).

by depositing the sample on one salt plate, placing another on top of it and then gluing this one down while the sample is pressed flat with a weight. Figure 15 shows a particle prepared in this manner. It does not look very elegant, but it works.

We found that liquids can be difficult to deal with if they are present as very minute droplets on a surface. If the droplets are in the range of 20 to 40 micrometers in diameter, they can sometimes be transferred to a salt plate by dropping the salt plate on the droplet and then pulling it away, hoping that sufficient sample has been transferred to the salt for analysis. This is often uncertain and usually ineffective.

We have been very successful with that type of problem using, what we call, a capillary brush. In Figure 16 we have three capillaries in various stages of manufacture. The upper capillary is 200 micrometers in inside diameter and is made of borosilicate glass. It is roughly 4" in length. The center capillary has a short length of borosilicate glass wool stuffed into the end. Using a hobbyists micro torch, the glass wool is fused to the inner surface of the capillary and then the excess wool is snipped at an angle close to the end of the capillary. The result is an open tube having a brush end which comes to the point as is shown in the bottom of Figure 16. After the capillaries are rinsed with solvent and dried, they are ready for use. Very small droplets of liquid can be swabbed up by the capillary action of these brushes and can then be transferred to a salt window. Sometimes it is necessary to prepare two brushes: one of 200 micrometers inside diameter and the other of 500 micrometers. The larger one is filled with solvent and dabbed at the sample. At the same time, the smaller brush is used to swab up the dissolved droplet. The solution in the small capillary can then be deposited on a salt window by gentle and repeated dabbing of the brush.

The droplet and deposit can be constrained to about a 200 micrometer circle or square. Depositing a solution from a capillary that does not have a brush tip, yields a very large droplet. The more solvent that is used, the larger the circles become. Most of the sample is in a ring at the outer edge of the circle with virtually nothing in the center.

One might ask, why not rinse the entire surface and then evaporate the solvent and transfer it to a salt plate. We have run into a problem in the past involving a computer disk that had extractable material in the surface coating as well as the contaminant droplets. The extractables accounted for so much of the sample that it is impossible to detect the small quantity of the contaminant in the spectra. Thus, by using this technique, one can avoid the problem of contamination from surrounding areas.

We frequently encounter polymers that have high levels of pigments and fillers. Samples such as paint chips and molded parts may have carbon black or other filler content. These fillers, either produce an energy loss or an

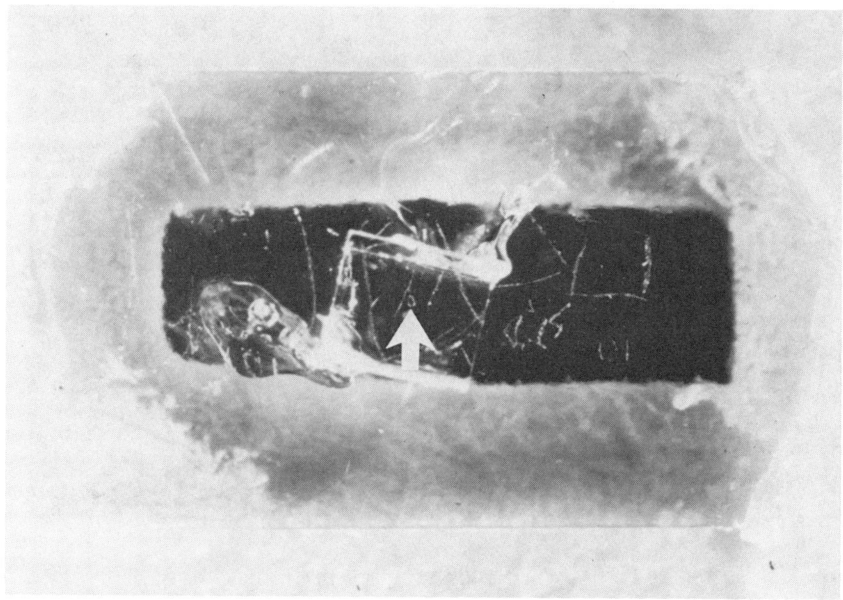

Figure 15. Elastomer specimen (arrow) pressed between two salt plates that have been glued together.

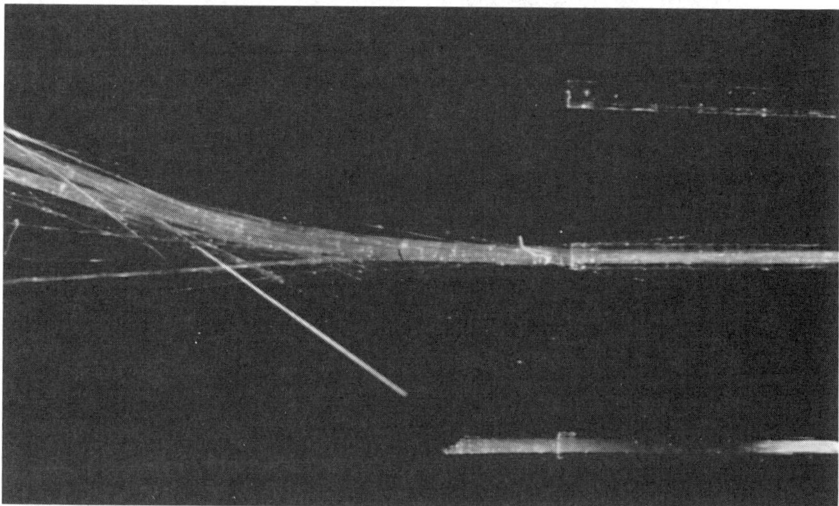

Figure 16. Capillary brush manufacture. Top: a 100 μm ID capillary tube. Center: Capillary with glass wool stuffed into one end. Bottom: Finished capillary brush.

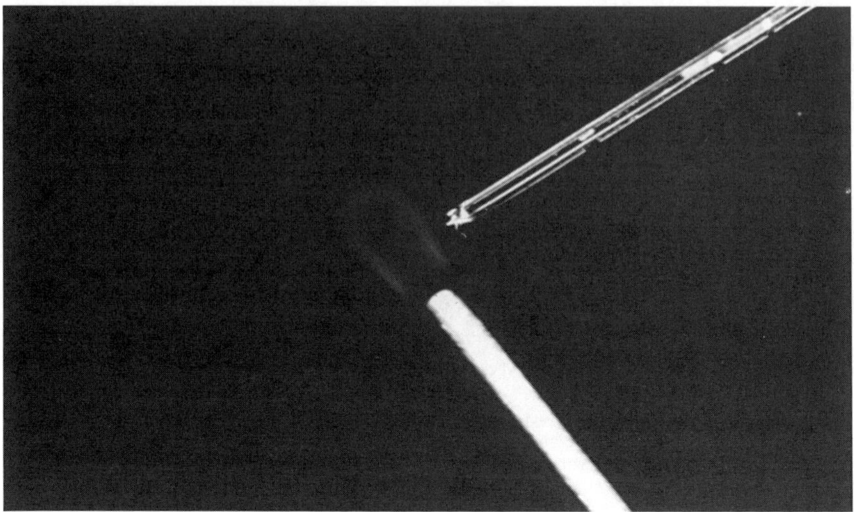

Figure 17. Specimen being sealed and pyrolyzed in capillary brush.

intolerable level of interference. Polymer pyrolysis or destructive distillation, followed by infrared analysis of the pyrolyzate, is a generally accepted technique for dealing with large samples of this type. But what do you do when the sample is very small? What we have done is to adapt the capillary brush technique for preparing and transferring pyrolyzates of very small samples.

Figure 17 shows the open end of a 200 micrometer capillary brush which has had inserted into it a specimen of filled and pigmented polymer. The sample has been pushed a few millimeters into the tube and then the end has been sealed with the micro torch shown in the figure. After the torch has been used to seal the end of the capillary, the flame is played up over the sample, pyrolyzing it. Tiny droplets of condensate accumulate a short distance from the ash. The tube is snapped off between the condensate and the ash, and a droplet of solvent is touched to the end of the tube, washing the pyrolyzate to the brush end. The dissolved pyrolyzate is then applied to the salt plate and analyzed.

Figure 18 shows triplicate analyses of portions of the same sample. You can see that the spectra are virtually identical, showing the high level of reproducibility. Incidentally, each of these samples was less than one microgram in weight, including a considerable amount of fillers and pigments. The reason we typically obtain such a high level of reproducibility - greater than that we sometimes get for larger samples - is because the pyrolysis takes place almost instantly with no local heating. A second point is that the total

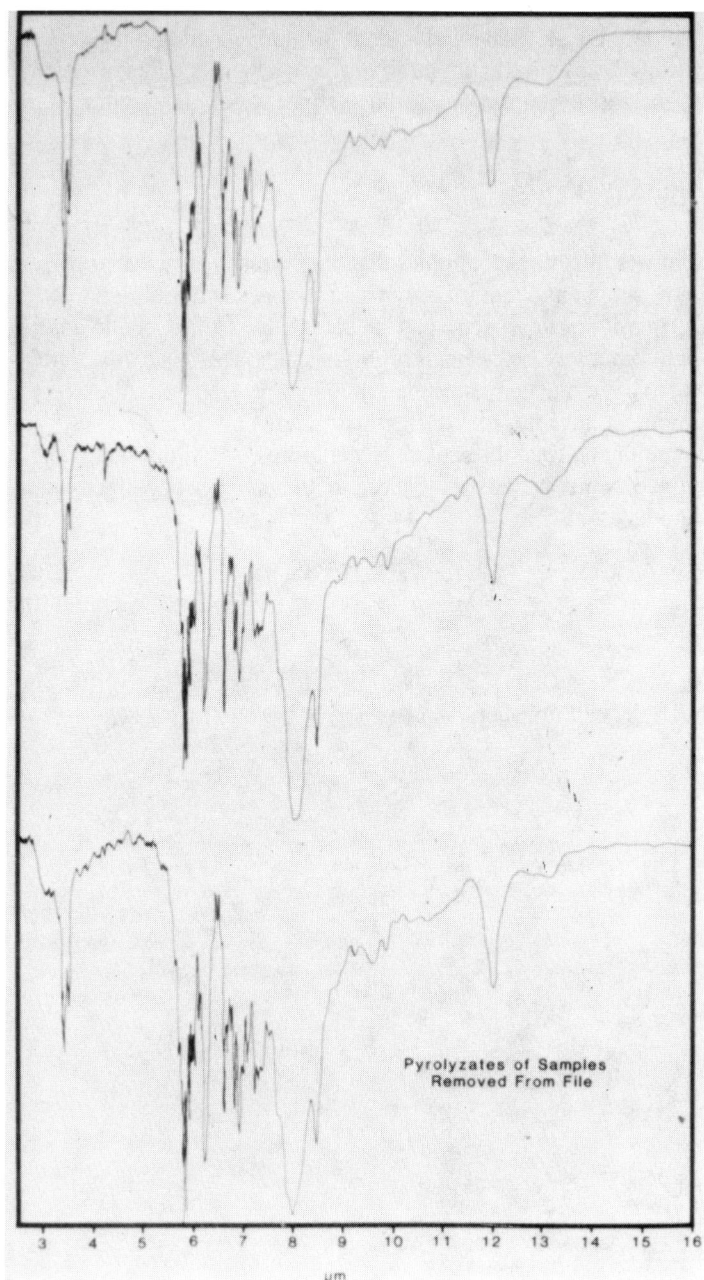

Figure 18. Triplicate analyses of sub-microgram specimens showing reproducibility.

sample is washed onto the salt plate and analyzed. When dealing with large pyrolyzates, we sometimes get fractionation of the condensate and since we may not analyze the entire sample, we can have slight variations in our spectra.

4.4 CONCLUSIONS

Some of the examples illustrate techniques that may already be familiar to you. They are relatively simple and require only basic supplies and tools, and yet they improve the quality of the spectra and, in some cases, allow one to analyze samples that might otherwise be difficult or impossible. This is an area that needs improvement. We must advance the art of sample preparation just as the designers of equipment have improved their instruments.

I leave you with one final thought. No matter how small the sample is that you can analyze, someone will come along with a smaller one that must absolutely have to be identified.

REFERENCES

1. McCrone, W. C. and Delly, J. G., *The Particle Atlas*, 2nd Ed., Ann Arbor Science Publishers, Ann Arbor, MI, Vol. 1: 225-227 (1973).

5

Polymer Composite Failure Analysis by Infrared Reflectance Microspectroscopy

DENNIS J. GERSON* and **CATHERINE A. CHESS**† *IBM Instruments Inc., Danbury, Connecticut*

5.1 INTRODUCTION

Fourier transform infrared (FT-IR) spectroscopy is universally recognized as an extremely important, effective and powerful tool for the characterization of polymeric materials [1]. Today, with the incorporation of composite materials into more and more products, the ability to characterize the composite polymeric materials is essential for the industrial analytical laboratory. Composite materials present several unique challanges for the industrial chemist and his repertoire of spectroscopy techniques. The materials are usually highly resistant to mechanical or physical stress, are more often than not optically opaque and may also contain large concentrations of inorganic or amorphous materials. Furthermore, these samples are typically available only in the final product form, making sample preparation very difficult [2]. The inability to prepare the samples for use with routine microsampling methods, such as micro-diffuse reflectance or micro-attenuated total reflectance, further complicates the problem for the analytical chemist [3].

**Present address, IBM Corp., National Engineering / Scientific Computing Center, 1503 LBJ, Dallas, Texas*

†*Present address, IBM Corp., National Service Division, 36 Apple Ridge Road, Danbury, Connecticut*

Recently, with the availability of high performance infrared microscopes, the analytical chemist is provided with an alternative to conventional microsampling methods. This alternative is *in situ* infrared reflectance microscopy. In this paper the use of infrared reflectance microspectroscopy to characterize polymer composite materials *in situ* will be discussed. Several examples will be presented which show the applicability of this technique to the analysis of failed composite materials as used in engineering and structural products.

5.2 EXPERIMENTAL DESIGN

The infrared spectrometer system used in these analyses consists of an IBM (Danbury, CT) IR/44 FT-IR Spectrometer and IBM Infrared Microscope. The FT-IR optical design provides a highly stable and reproducible infrared beam, via a rectangular design air bearing drive interferometer, closed circulation water cooled source, high thermal mass optical bench and white light correlation interferometer [4]. The microscope design, optimized for reflectance microscopy, utilizes high efficiency Cassegrainian reflecting optics and a matched focus, small element high sensitivity MCT detector to provide the optimum senstivity and reproducibility for the analysis [3].

The materials analyzed by infrared reflectance microspectroscopy were obtained from commercially available sources and utilized without sample preparation, except for mounting on the microscope stage. The spectra are the result of up to 10 minutes of data acqusition at 4 or 8 cm^{-1} resolution. The experimental parameters and optical accessories for the microscope were optimized for each sample [3]. These are summarized in Table I. The optical diagrams for the microscope and reflectance objectives are shown in Figures 1 and 2, respectively.

5.3 RESULTS AND DISCUSSION

5.3.1 Engineering Seals

Composite materials consisting of graphite or Vespel (TM) are currently being used as engineering seals in elevated pressure or elevated temperature applications because of the materials inertness to other chemical substances. An example of such use are the pump seals associated with a high pressure

Table 1. Experimental Parameters and Accessories

Parameter	Engineering Seal	Graphite Composite
Acquisition Time	5 mins	2 mins
Resolution	8 cm^{-1}	8 cm^{-1}
Correlation Mode	White Light	White Light
Objective (Power)	15x	36x
Aperture	150 micron	250 micron
Relative Reflectivity (% of Background)	30%	10%

liquid chromatograph (HPLC) pump. In this application, the graphite-teflon seals were observed to fail during the early stages of pump operation. The analytical problem was to determine what was causing the seals to fail and if this cause was chemical or mechanical in nature. Under microscopic examination by visual reflectance microscopy several shiny streaks were observed over the surface of the seal. A schematic representation of a failed pump seal is shown in Figure 3.

The spectrum of a good, non-streaked, area of the seal is shown in Figure 4a. The spectrum exhibits strong *reststrahlen* bands in the region of the Teflon® absorbance near 1200 cm^{-1}, and is typical of the reflection spectrum of graphite-fluoropolymer materials. The comparison of this spectrum to that of a good pump seal, using the spectral search commands, gave a very good match,

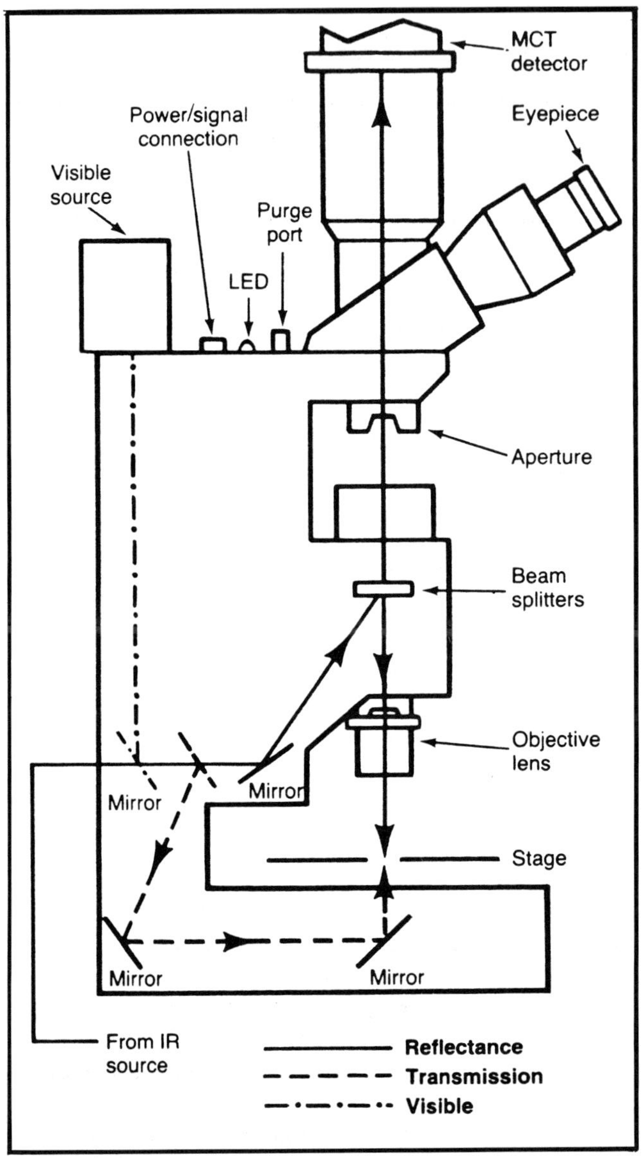

Figure 1. Optical system for infrared microscope.

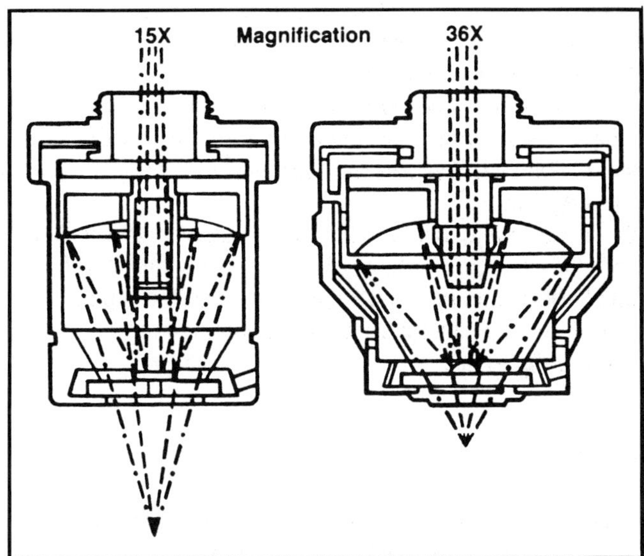

Figure 2. Optical diagram of reflectance objectives.

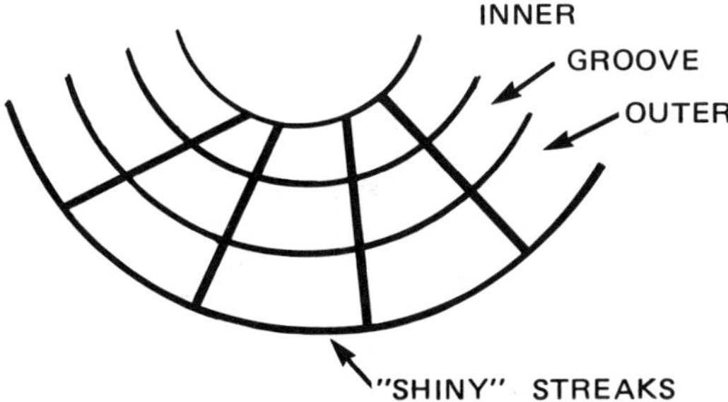

Figure 3. Schematic representation of failed pump seal.

indicating that the non-failed areas of the seal are the same as a good seal. Analysis of the difference between the spectrum of the good area (Figure 4a) and that of the failed, "shiny streaked" area (Figure 4b), showed some differences. The difference spectrum, the result of computer assisted automatic subtraction and noise-adaptive smoothing, is shown in Figure 4b. Performing a search of the IBM Standard Polymer Reference Library gave an excellent match to the reflectance spectrum of Teflon.

From these results, it was concluded that the failed pump seals were not chemically different from the good pump seals. The defective areas on the failed seals showed more Teflon than the good areas. This result is consistant with a mechanical failure of the seals and not a chemical failure due to corrosion or decompositon of the seals.

5.3.2 Structural Composite

Polymer resin coated graphite fiber mats are being used in manufacturing as light weight substitutes for structural steel and non-alloy metals. The mats, when cured, exhibit structural strengths superior to steel and have the added

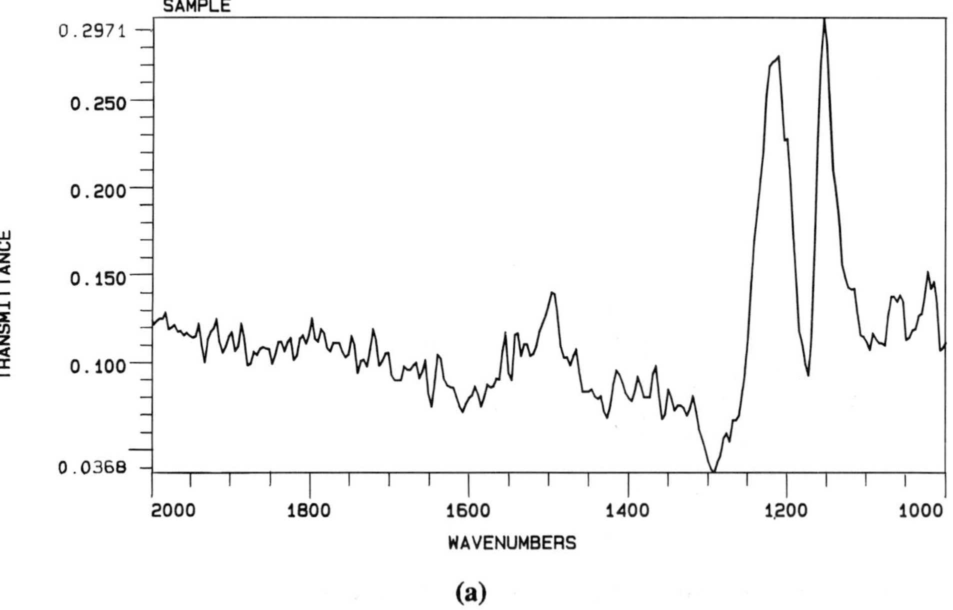

(a)

Figure 4. Spectra of the pump seal; **(a)** good area, **(b)** failed area, **(c)** difference.

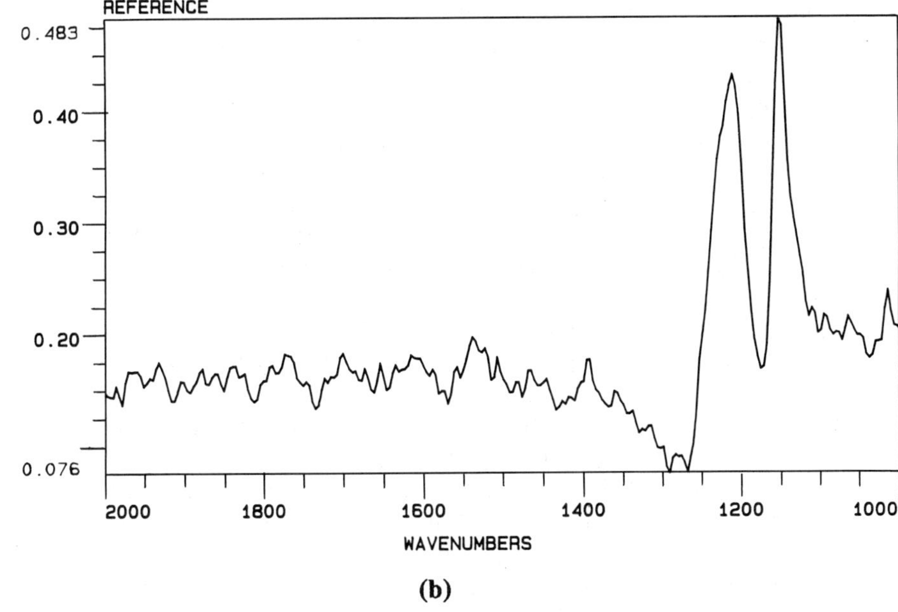

(b)

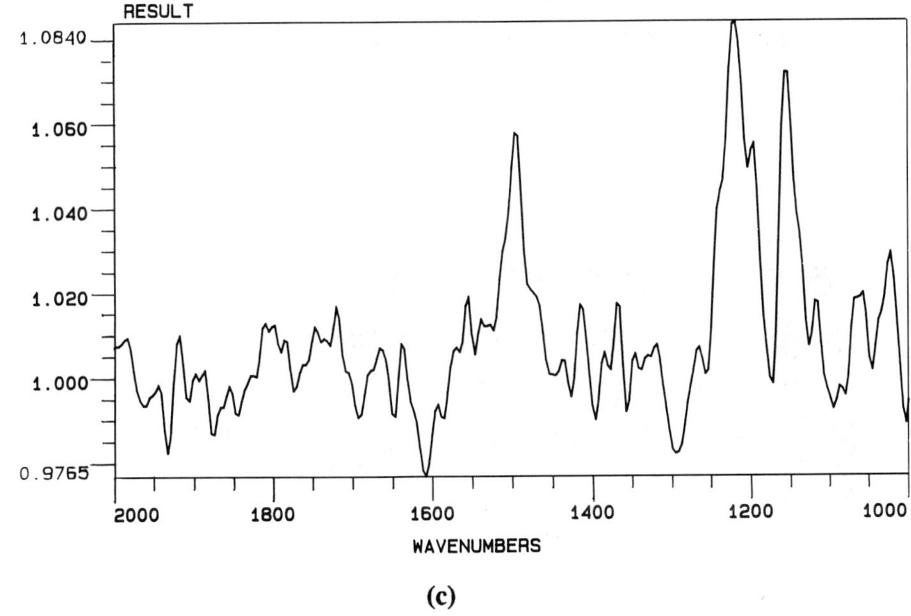

(c)

Figure 4. (continued)

advantage of being able to be hardened into any shape using a mold. Current uses of these type of materials include helicopter rotors, airplane bodies and automobile bumpers. In this case, the characterization of a failed structural replacement was performed using reflectance microspectroscopy. The problem was to determine why the part failed structurally and how the degree of hardening or curing can be monitored spectroscopically.

The graphite composite mat consists of two fiber layers and an overcoating of polymer resin. A cross-sectional representation of the mat is shown in Figure 5. The spectrum of the uncured graphite mat is shown in Figure 6a. The spectrum exhibits a non-linear baseline due to the highly scattering nature of the carbon fiber matrix. The spectral features observed riding the baseline are from the resin. By applying a high order polynomial baseline correction to the spectrum the scattering baseline from the carbon fibers can be eliminated [5]. The result of an automatic computer-controlled 5th order baseline correction is shown in Figure 6b. The data has been intensity normalized using a Kubelka Munk function. This function corrects for non-linear diffuse scattering from the sample [6]. Spectra of a failed and an acceptable sample were obtained using the same procedure. Upon examination of these spectra, shown as an overlay plot in Figure 7, reveals a difference in the intensity of a very strong absorbance at 1520 cm^{-1}. Futher comparison of these spectra with that of the fluid resin, washed from the sample and prepared as a cast film (Figure 8) reveals the intensity of the band at 1520 cm^{-1} to be of the same relative intensity in the failed sample as in the resin [7].

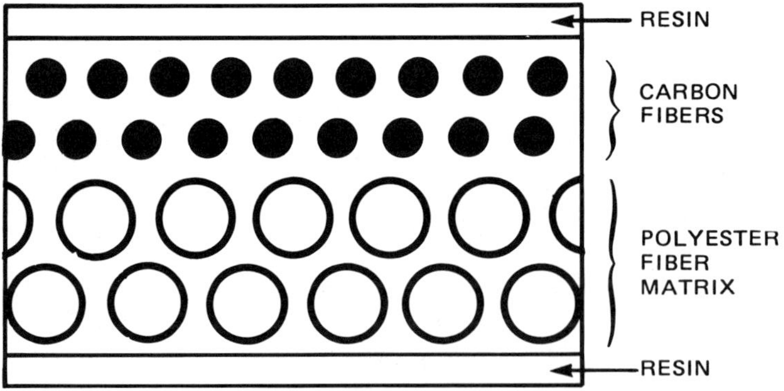

Figure 5. Schematic representation of composite mat.

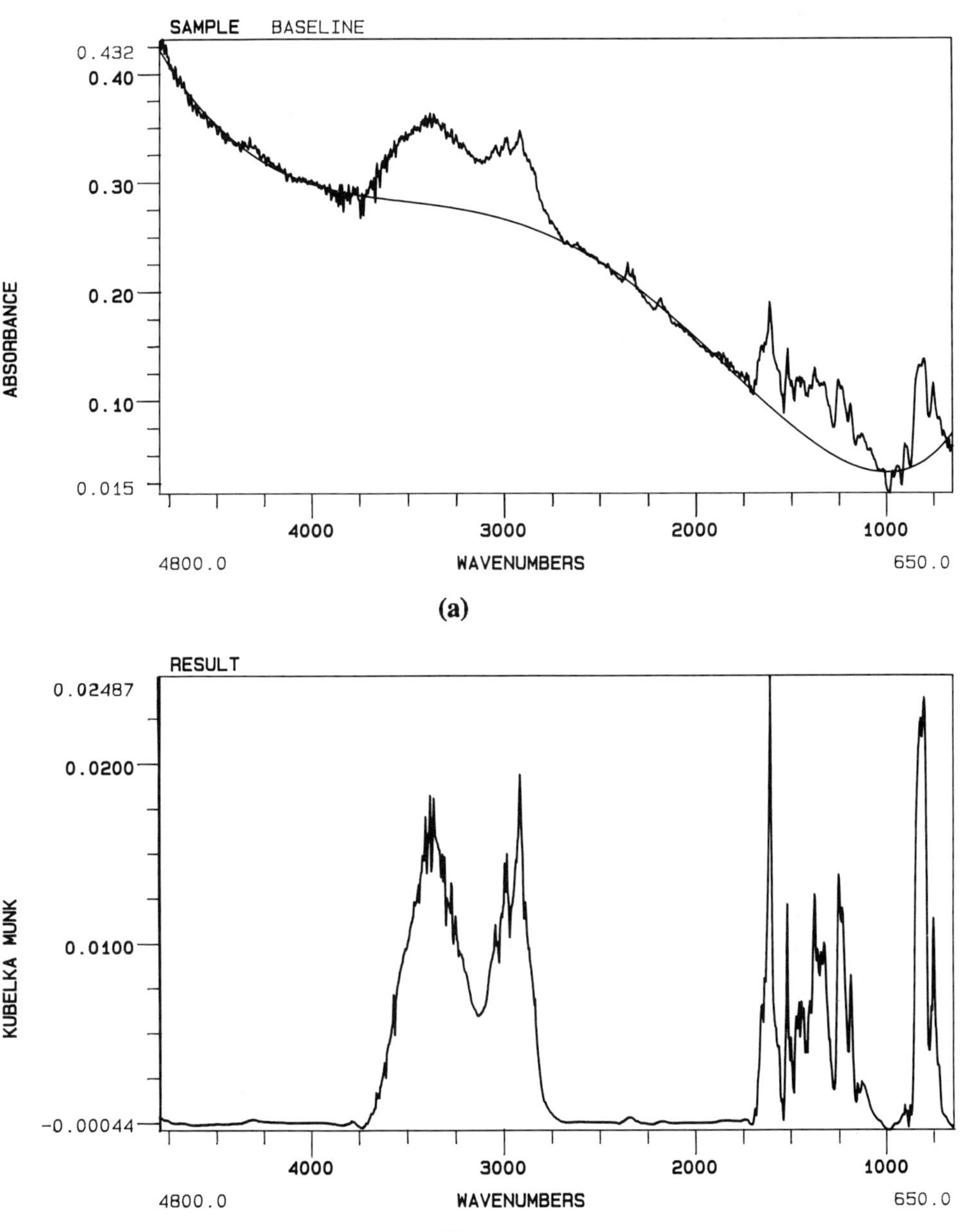

Figure 6. Spectra of uncured composite mat; **(a)** raw data, **(b)** after chemometrics.

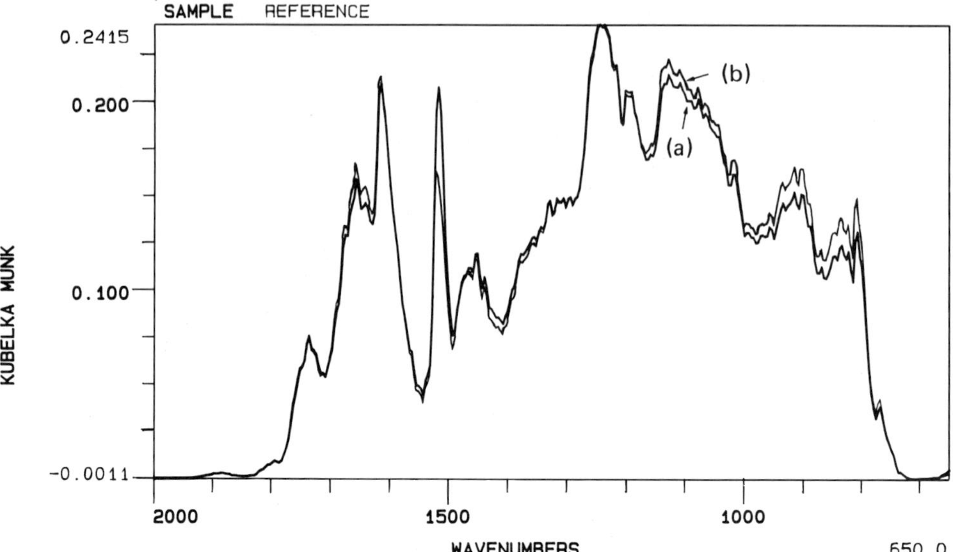

Figure 7. Overlay plot of the spectra of composite mat after application of chemometric methods: (a) sample, cured; (b) reference, uncured.

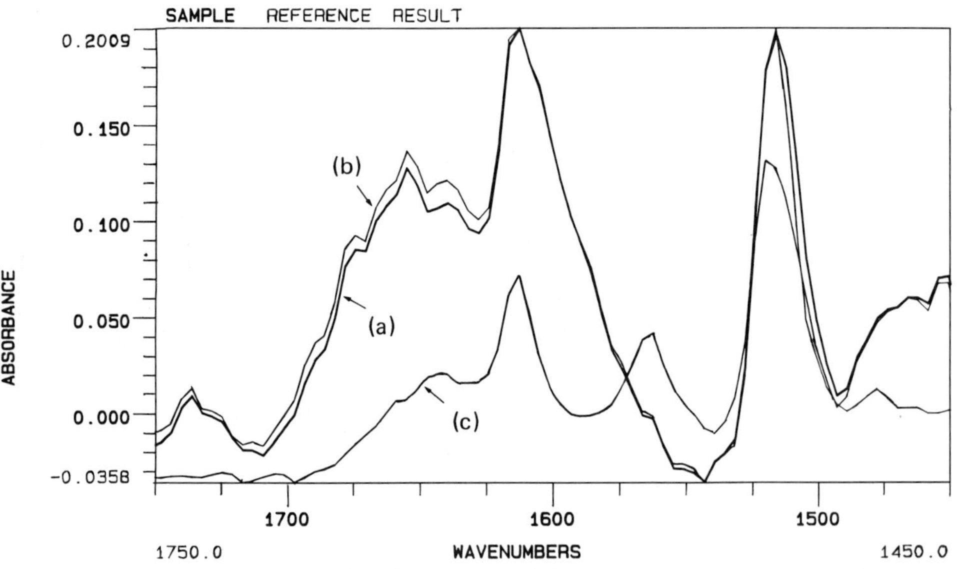

Figure 8. Overlay plot of the spectra of the composite mat and extracted resin; (a) sample, cured; (b) reference, uncured; (c) result, resin.

From the above data, it is clear that the failed material contained excess fluid or uncured resin. The degree of hardening of the material can be monitored by comparing the relative absorbance of the 1520 cm^{-1} band in a test sample to that of a good finished product. Application of this method in our lab has shown that the material cures to an acceptable hardness after 24 hours at 90 °C. The material cures to 75% of the acceptable level after 96 hours at room temperature.

5.4 SUMMARY

It has been shown that in order to obtain good quality microspectroscopic data from polymer composite samples *in situ*, a microscopic system containing high efficiency reflectance optics is extremely useful. The spectrometer data system should contain the chemometric methods needed to correct for the nonlinear baselines and low-signal-to noise inherent in the observed spectra for these samples.

ACKNOWLEDGEMENTS

The authors would like to acknowledge the the work of Drs. D. Sparks (Nicolet Instrument Corp.) and R. Lam (IBM Corp.) for the development of the chemometric methods utilized in this project. The authors would also like to thank Dr. J. Casper (IBM Corp.) and Mr. F. Hewitt (American Optical) for their support and encouragement during the course of this project.

REFERENCES

1. P. R. Young and A. C. Chang, *SAMPE Quarterly*, 17: 32-39 (1986).
2. J. C. Shearer, D. C. Peters and T. A. Kubic, *Trends Anal. Chem.*, 4: 246-251 (1985).
3. S. L. Smith, *Research and Development*, 28(9): 113-118 (1986).
4. A. J. Rein and K. S. Morris, *American Lab.*, 18(9): 86-97 (1986).
5. *IBM Journal of Technical Disclosure*, 55(5): 18 (1987).
6. P. R. Griffiths, *Applications of FTIR Methods*, in *Proc. NATO ASI #27* (1981).
7. The resin in question is an epoxy-amine in butyl alcohol.

6

Simultaneous FT-IR Microspectroscopy and Differential Scanning Calorimetry: Polymer Studies

FRANCIS M. MIRABELLA *USI Chemical Company, Rolling Meadows, Illinois*

6.1 INTRODUCTION

A novel technique has been previously demonstrated for the simultaneous recording of differential scanning calorimetric (DSC) data and infrared spectra recorded in a FT-IR microspectroscopy accessory (IRMA) on specimens of microscopic dimensions [1]. The experimental apparatus and operational variables of this technique have been further refined. The purpose of this report is to describe these refinements and to present experimental data on some applications of this technique to polymer systems.

6.2 EXPERIMENTAL

The infrared spectrophotometer used is a Nicolet (Madison, WI) 6000 FT-IR. All spectra were recorded at 4 cm $^{-1}$ resolution. Acceptable S/N was obtained by the coadding of from 10 to 100 spectra at a mirror velocity of 0.586 cm/s. A Digilab (Cambridge, MA) transmittance infrared microscope with 250 μm MCT detector was used to focus the infrared radiation through the DSC cell. The DSC cell used was a Mettler (Hightstown, NJ) FP84 thermal analysis

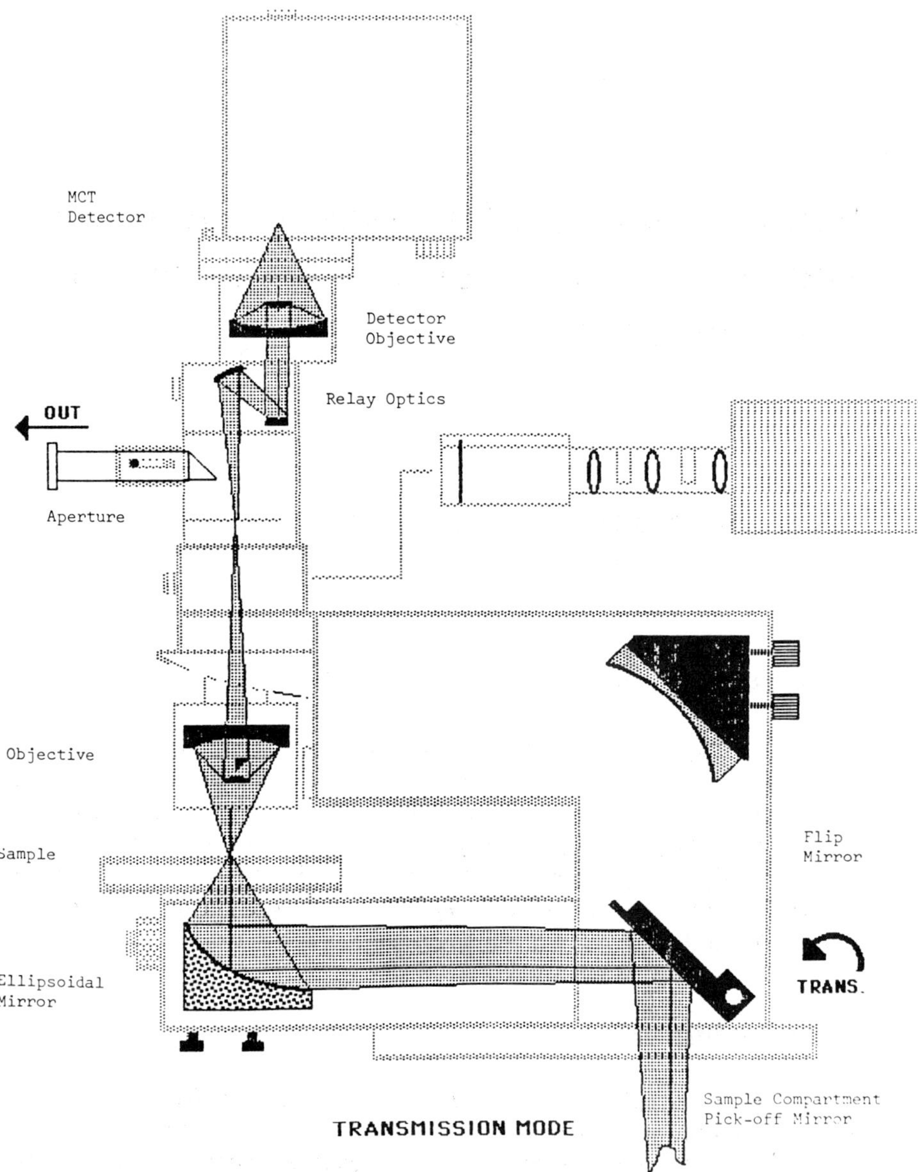

Figure 1. Schematic Diagram of the FT-IR microsampling accessory.

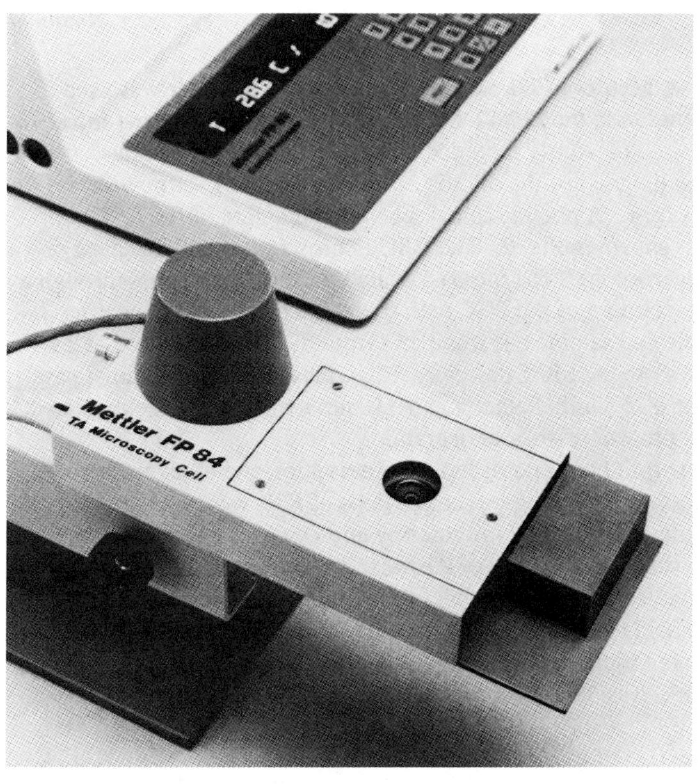

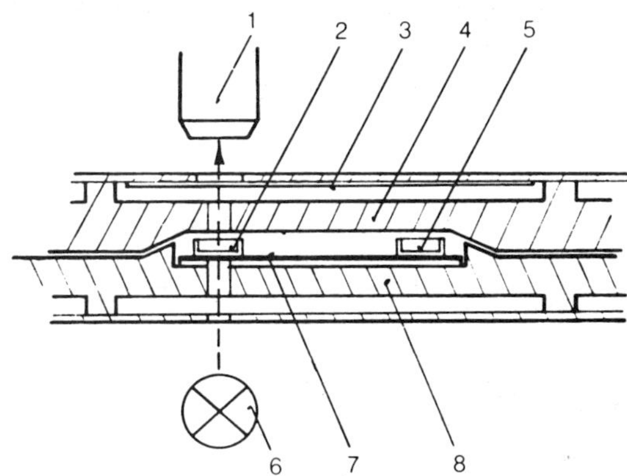

1 Microscope objective
2 Sapphire sample crucible
3 Heat protection filter
4 Metal plate with heating wires
5 Sapphire reference crucible
6 Microscope light source
7 DSC measuring sensor
8 Metal plate with heating wires and Pt100 resistance sensor

Figure 2. Photograph and schematic diagram of microscopy/DSC cell.

microscopy cell. A Mettler FP80 central processor with recorder was used to control the temperature in the FP84 DSC cell. DSC thermograms and infrared spectra could be simultaneously recorded with this combined apparatus.

A schematic diagram of the microspectrometer with the infrared beam path is shown in Fig. 1. A photographed schematic diagram of the DSC microscopy cell is shown in Fig. 2. The DSC microscopy cell was placed on the stage of the microscope. The infrared beam was required to pass through a 2.5 mm hole in the bottom of the DSC cell, travel 3.0 mm (thereby passing through the sample and sample cup), and exit from the DSC cell through a 3.0 mm hole on the way to the MCT detector. Thus, the infrared beam must pass through a cylinder of 2.5 mm diameter and 3.0 mm length. This requires a careful alignment procedure prior to operation.

In order to permit both microscopical observation and transmission infrared spectroscopy of the sample, sample cups of KBr were used to hold the sample. For simultaneous optical microscopy and DSC, sapphire sample cups are used. The thermal conductivity of KBr is 0.7 g-cal/s-cm-°C, that of sapphire is 0.6 g-cal/s-cm-°C, while that of aluminum is 55.9 g-cal/s-cm-°C, all at 100°C. Thus, KBr is a reasonably good conductor of heat, but it is also transparent to visible light and infrared radiation in the wavelength range of 28,000 to 600 cm^{-1}. It is, therefore, an excellent material for use in the present application.

The construction of the KBr sample cups is shown in Figure 3. A trick which enabled the recording of sufficiently intense DSC thermograms, when the specimen being studied was very small, was to place an identical specimen in the reference cup but of larger mass. Then an inverted thermogram was obtained of larger intensity representative of the difference of the reference minus the sample mass.

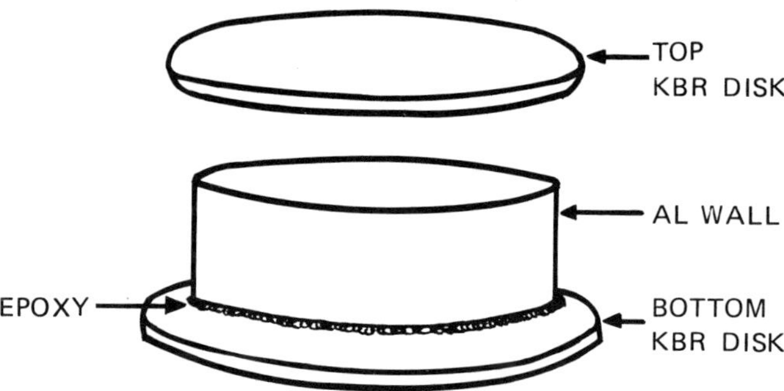

Figure 3. Transparent KBr and aluminum DSC cup.

6.3 DISCUSSION

The first application of the simultaneous FT-IR microspectroscopy and DSC technique in this work was to the study of the effect of temperature on the intensity of infrared absorptions. The infrared spectra of high density polyethylene, low density polyethylene, linear low density polyethylene and polypropylene were studied as a function of temperature. The polymers were studied as they were heated from room temperature through and above the melting point and then cooled through and below the freezing (crystallization) point. Many spectra were obtained as the temperature was changed and then the most informative spectra were chosen which exhibited structural changes of the polymer. If changes were occurring slowly 100 spectra were coadded but if changes were occurring rapidly, e.g., near the melting or freezing transitions, only 10 spectra were coadded to yield the averaged spectrum. The purpose here is to suggest the possibilities of this technique. Therefore, these data will not all be considered in depth. However, by way of example, Figure 4 shows the changes in the methyl (1378 cm^{-1}) and methylene (1368 cm^{-1}) region of the infrared as low density polyethylene is melted. The DSC thermogram is also shown and was typical of low density polyethylene melting behavior and could be used to correlate with the changes in the IR spectra observed. Melting occurs at a peak temperature of about 112°. Recrystallization exhibited a reverse behavior of the IR spectra. Studies such as these can be quite informative in following the chain dynamics as polymers are heated and cooled through transitions. The DSC response can be increased in intensity by placing a larger (identical) sample in the reference pan than the sample on the sample pan. This yields a more intense DSC curve than would be obtained from the sample being directly observed, but in the reverse direction (i. e. an endotherm becomes an exotherm).

The next application was to that of polymer degradation. The degradation of a small specimen of poly (vinyl alcohol) was followed as it was heated in air at 175°C over a period of 3 hours. The major change was a loss of hydroxyl groups (about 3400 cm^{-1}) and a rapid increase in ketonic carbonyl groups (about 1710 cm^{-1}) as the degradation proceeds. This can be seen in Figure 5 in which the degradation can be seen to proceed markedly between 30 minutes and 3 hours. The specimen was about 0.01 mm (0.4 mil) thick. Again, the rather long DSC thermogram is not shown but could be used to correlate with the IR spectra. It would be interesting to follow such degradations of polymers in the FT-IR microsampling accessory in multilayer films in which the polymers are in contact with one another. It has been shown that such multilayer films could be qualitatively identified using the FT-IR microsampling technique [2]. Thus, the simultaneous FT-IR microsampling and DSC technique might prove useful to determine the effect of the close proximity of the polymeric layers on one another as they are subjected to some thermal treatment.

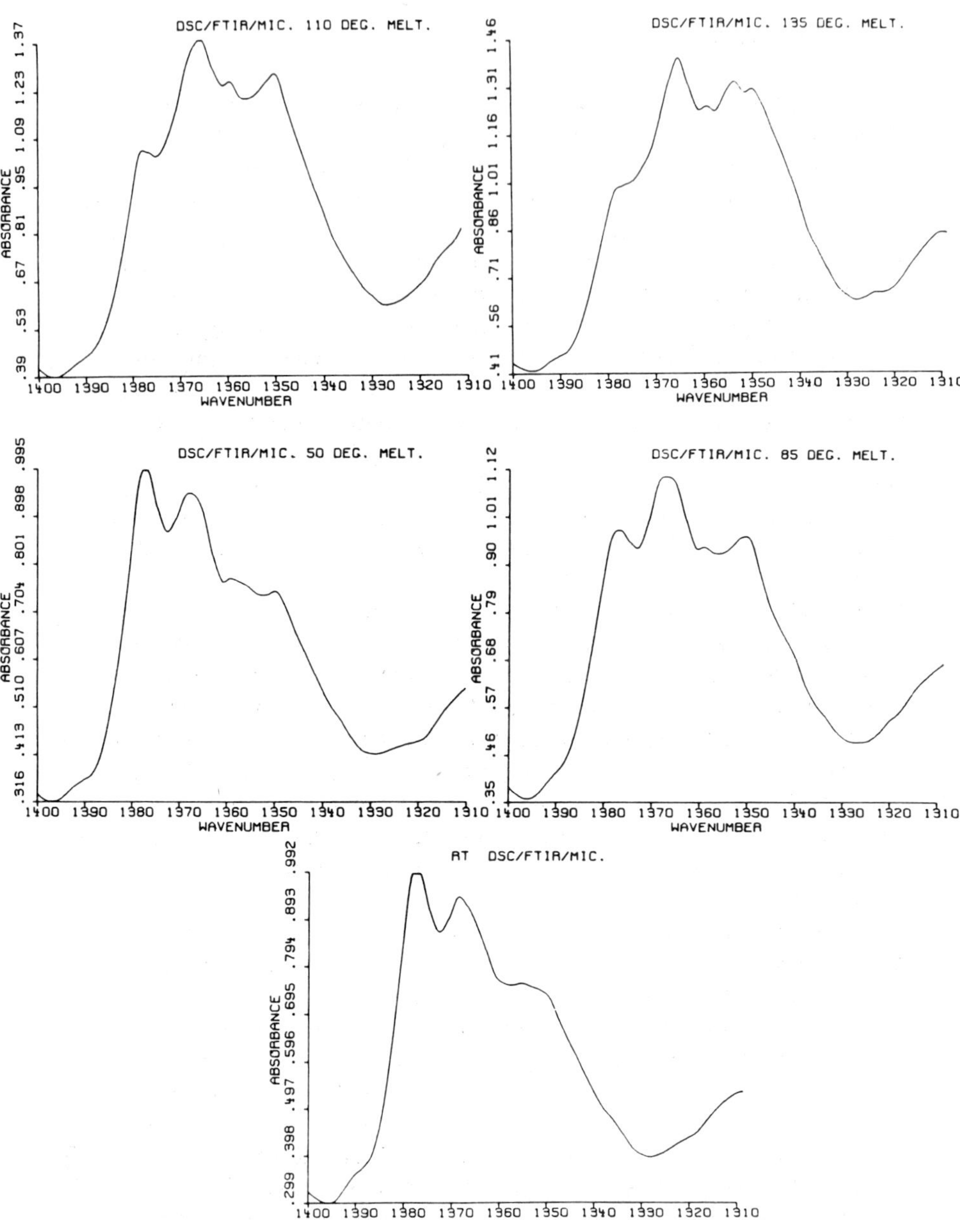

Figure 4. Selected infrared spectra of the melting of a low density polyethylene and the DSC melting endotherm.

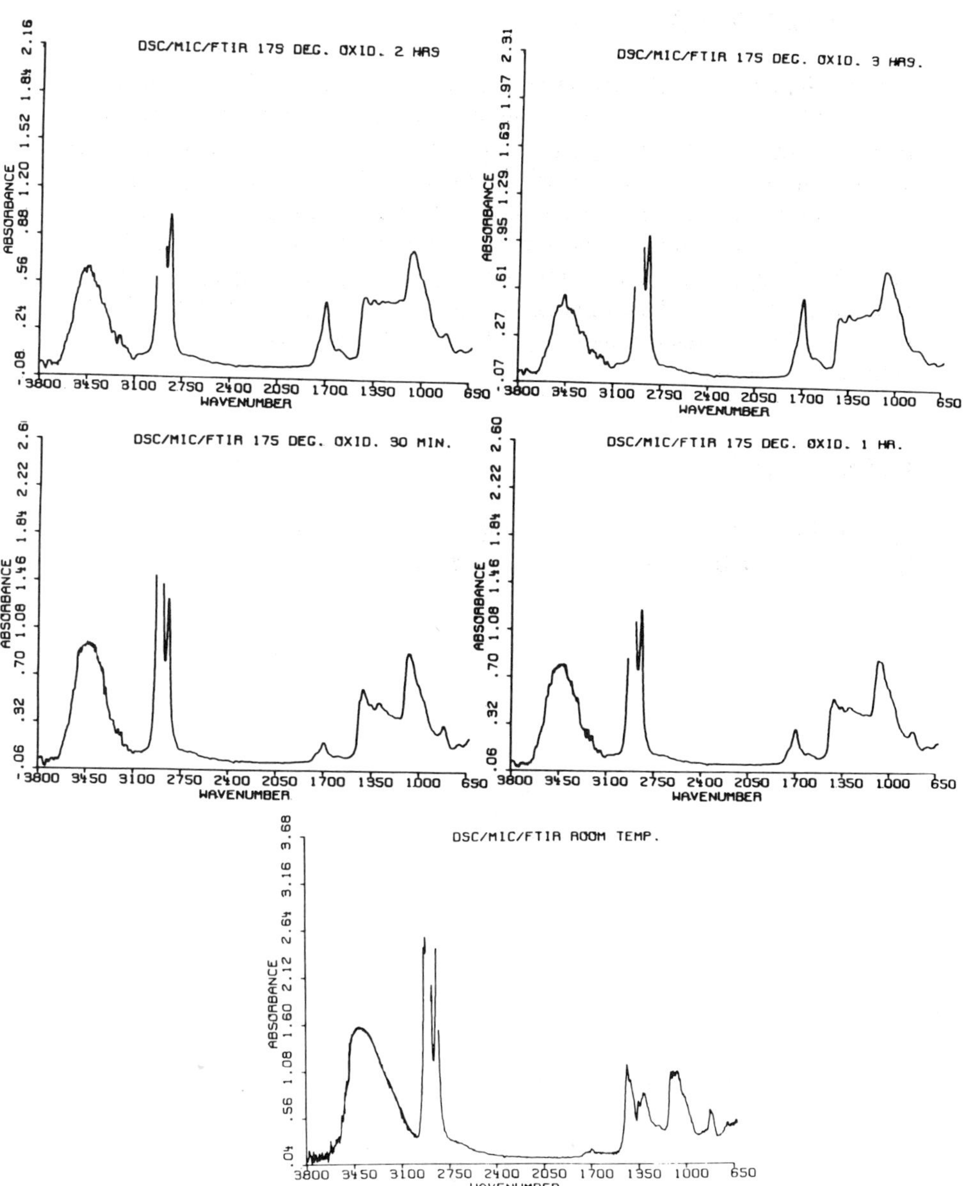

Figure 5. Selected infrared spectra of the degradation of a poly (ethylene vinyl alcohol) in air at 175°C.

6.4 CONCLUSIONS

The simultaneous FT-IR microsampling and DSC technique has been shown to be useful for studying the effect of thermal treatments on the infrared spectra of polymers and for following changes in polymer structure, such as degradation. The atmospheric gas may be changed in these studies. Microtomed cross-sections of multi-layer films can be studied by this technique. The adjoining polymer layers, containing contaminants, metal residues and catalyst residues might be found to affect the degradation, crystallinity, etc. of the polymeric layer being studied in these multilayer systems. The new generation of FT-IR microsampling accessories will no doubt facilitate advancements in the demonstrated technique.

REFERENCES

1. F. M. Mirabella, Jr., *Appl. Spectry.*, 40: 417 (1986).
2. F. M. Mirabella, Jr., *Polym. Eng. Sci.*, 26: 605 (1986).

Applications of Polarized Infrared Microspectroscopy

7

Dichroic Infrared Spectroscopy with a Microscope

BRUCE CHASE *E. I. DuPont de Nemours & Company, Central Research Department, Wilmington, Delaware*

7.1 INTRODUCTION

One of the continuing problems never properly addressed by infrared spectroscopists, is the determination of molecular orientation in single fibers. Such orientation information can be readily obtained from polarized infrared spectroscopy, if the signal to noise ratio is sufficient and the polarization purity is good. The polarization purity is defined as the amount of parallel polarized light transmitted when the polarizer is set to pass perpendicularly polarized light. If the orientation could be measured as a function of spinning speed, draw ratio or other processing parameters, then the ultimate physical properties of the fiber might be understood in terms of molecular characteristics. These types of studies have been attempted with fiber bundles and transmission spectroscopy, but as light passes through the bundle, there are regions which are optically dense (seven or eight fiber diameters thick) and there are regions which are totally transparent, where there are no fibers at all. This results in a tremendous stray light problem. Similar studies have been tried using ATR (attenuated total reflectance) and other approaches, but clearly, the infrared microscope offers the most potential for obtaining useful information on this problem.

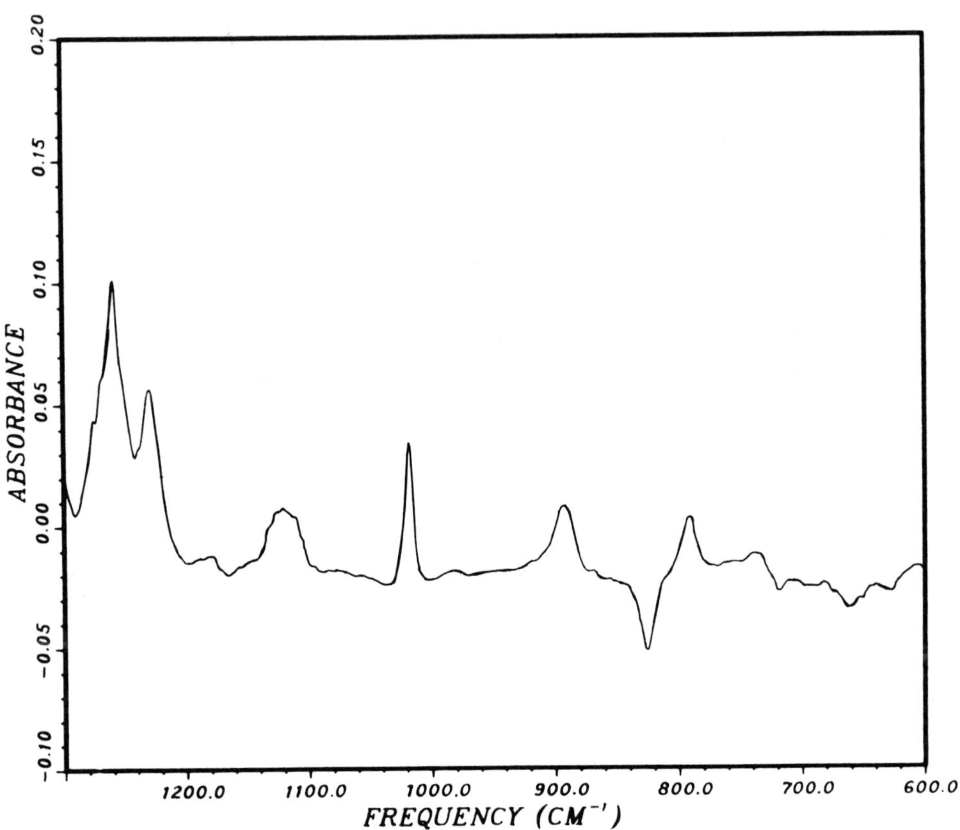

Figure 1. Partial dichroic spectrum of poly (p-phenylene terephthalamide) ribbon recorded in a conventional transmission experiment.

7.2 EXPERIMENTAL

Two different microscope-interferometer configurations were used in this study of the effects of stray light on polarized spectroscopy of single fibers. One instrument had a significantly higher stray light level, thus allowing a comparison of the dichroic ratios obtained under different conditions. An IRMA microscope supplied by Digilab (Cambridge, MA), equipped with a 250 micrometer Infrared Associates (Cranbury, NJ) MCT detector was coupled to a Nicolet (Madison, WI) 20DX interferometer. This instrument was equipped for transmission studies with a single aperture for resolving spatial elements.

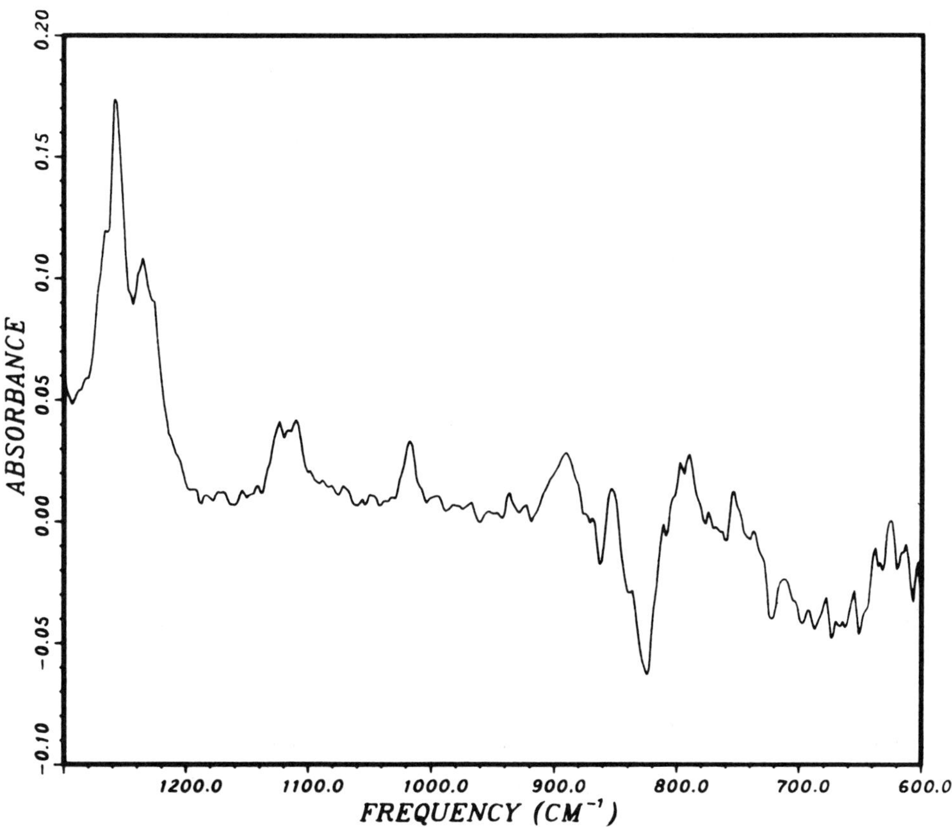

Figure 2. Partial dichroic spectrum of poly (p-phenylene terephthalamide) ribbon recorded in the infrared microscope (aperture 10μ x 120μ).

Spectra were recorded at 4 cm^{-1} resolution,and two times zero filling was employed; 512 scans were coadded for signal-to-noise enhancement. All data processing was done on a Digital Equipment Corporation (Maynard, MA) VAX 11/750. The second microscope was a Spectra-Tech (Stamford, CT) IR-plan with a similar 250 micrometer detector coupled to a Nicolet 20SX. This instrument was used courtesy of Spectra-Tech. Data were recorded under the same conditions as above. In all cases wire grid polarizers were used. Most spectra presented here are partial dichroic spectra, $A_{\perp}$ minus $A_{\parallel}$. See equation 1 for the definition of the true dichroic ratio.

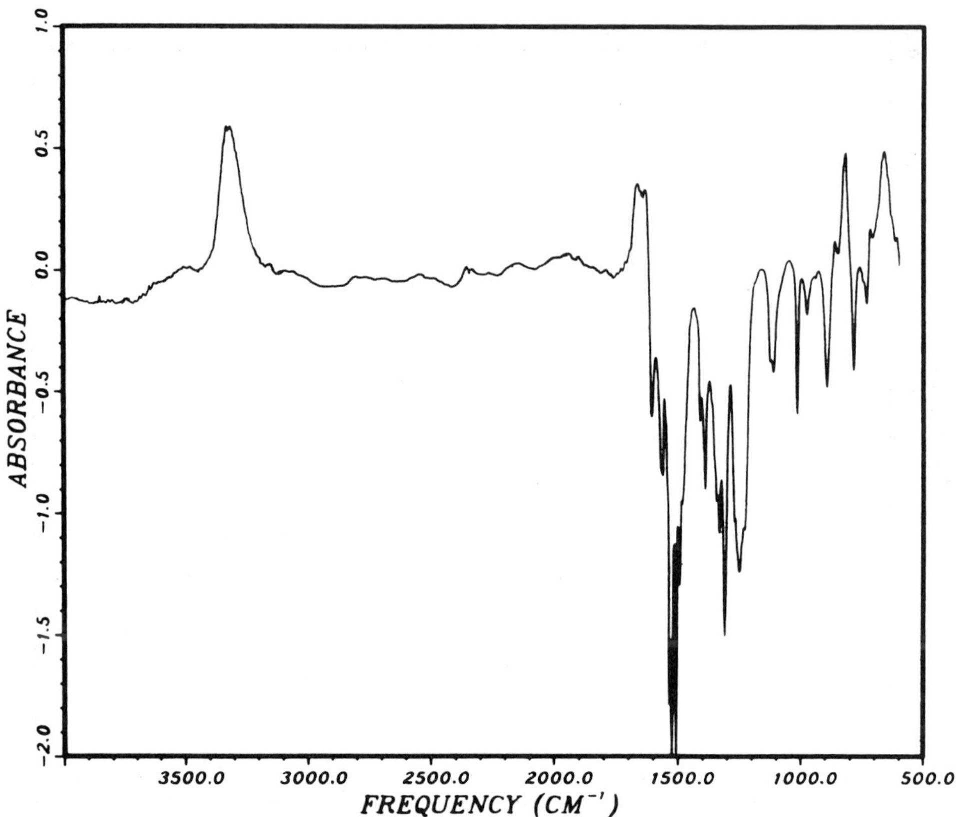

Figure 3. Partial dichroic spectrum of poly (p-phenylene terephthalamide) fiber recorded in the infrared microscope (aperture 10μ x 120μ).

7.3 RESULTS

The first difficulty that one must deal with in infrared dichroism, whether macroscopic or microscopic, is polarization scrambling. Does the instrument itself, independent of the sample, perturb the polarization? The second difficulty involves stray light. Since stray light can drastically decrease the photometric accuracy of any spectrum, especially for samples with strong absorptions, the perturbation on the measured absorbance as a function of the dichroic angle can well affect the calculated dichroic ratio.

The problem of polarization scrambling is not severe in the current generation of infrared microscopes. If an additional wire grid polarizer (Harrick Scientific, Ossining, NY) is placed at the sampling point and the transmission spectra are recorded for parallel and perpendicular orientations (with respect to fiber axis), the deviations from 100% and 0% transmission are about 20% and 5% respectively. The decrease in the 100% line is due to reflection losses and obscuration by the wires. The increase in transmission over the 0% value is probably due to some small degree of polarization scrambling or incomplete polarization by the wire grid element due to the high degree of focussing found with a low f/number beam at the sampling point.

In order to establish the validity of the dichroic measurement in the microscope, independent of stray light contributions, a sample whose area is large in comparision to the aperture being used must be measured. A ribbon of poly (p-phenylene terephthalamide), 5 mm in width, was used. Figure 1 is the partial dichroic spectrum of this sample recorded in a conventional transmission experiment on a Nicolet 7199 interferometer. The polarization purity of this instrument was established as better than 2 %; i.e. there is negligible scrambling. Figure 2 is the spectrum of the same ribbon recorded using the

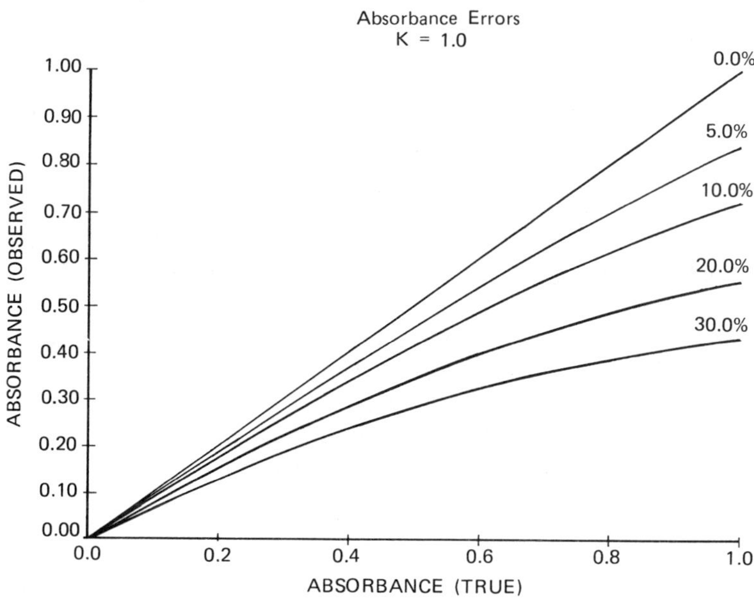

Figure 4. Absorbance error for a band of O.D.=1.0 at varying stray light levels.

microscope. There is quite good agreement between the two partial dichroic spectra, even with respect to the absolute values of the measured absorbances. Therefore, in the absence of stray light, the infrared microscope can produce dichroic data comparable to a conventional transmission experiment.

When fibers are examined, the problem of stray light is introduced since the sample size is now of the same order as the aperture size. However, good, qualitative results can still be obtained. Figure 3 is the partial dichroic spectrum of a twelve micron diameter fiber of poly (p-phenylene terephthalamide). The band at 3300 cm^{-1} is due to an N-H stretch. The amide N-H groups orient perpendicularly to the fiber axis. The band at 1550 cm^{-1} is due to a phenyl ring mode in which the dipole moment change is oriented along the para -para direction of the ring. This band should exhibit orientation along the fiber axis, which it does.

All of the results so far were obtained on a single aperture system. The remaining data were obtained on a Spectra-Tech IR-plan system. Wire grid polarizers were used and the spectrometer was a Nicolet 20DX. This instrument produces the same qualitative results as found for the singly apertured system, but with a reduced stray light level due to the use of Redundant Aperturing™.

Using either system, one can obtain partial dichroic spectra, but this information is only qualitative. Ideally, the true dichroic ratio, a dimensionless quantity which varies from 1 to -1, defined as

$$D.R. = A_{\parallel} - A_{\perp} / A_{\parallel} + A_{\perp} \quad [1]$$

should be determined. From the value of the dichroic ratio, ß, which is the angle between the reference axis (fiber axis) and the dipole moment vector can be determined. A value of 1.0 for the dichroic ratio would imply that the dipole moment vector and the fiber axis are parallel, while a value of -1.0 would mean they are perpendicular. Involved in this is the assumption that the fiber has radial symmetry. Initially, one would assume that if the absorbance values are incorrect due to stray light effects, then the dichroic ratio will also be in error and any determination of ß will be wrong. In order to determine how large this error might be, several simulations were done.

The expression for absorbance for both parallel and perpendicular polarizations in the presence of stray light is given in equation 2.

$$A_{\parallel} = \log\left\{\frac{1}{\varepsilon + (1-\varepsilon) * 10^{-K'\cos^2(\beta)}}\right\} \qquad A_{\perp} = \log\left\{\frac{1}{\varepsilon + (1-\varepsilon) * 10^{-K'\sin^2(\beta)}}\right\} \quad [2]$$

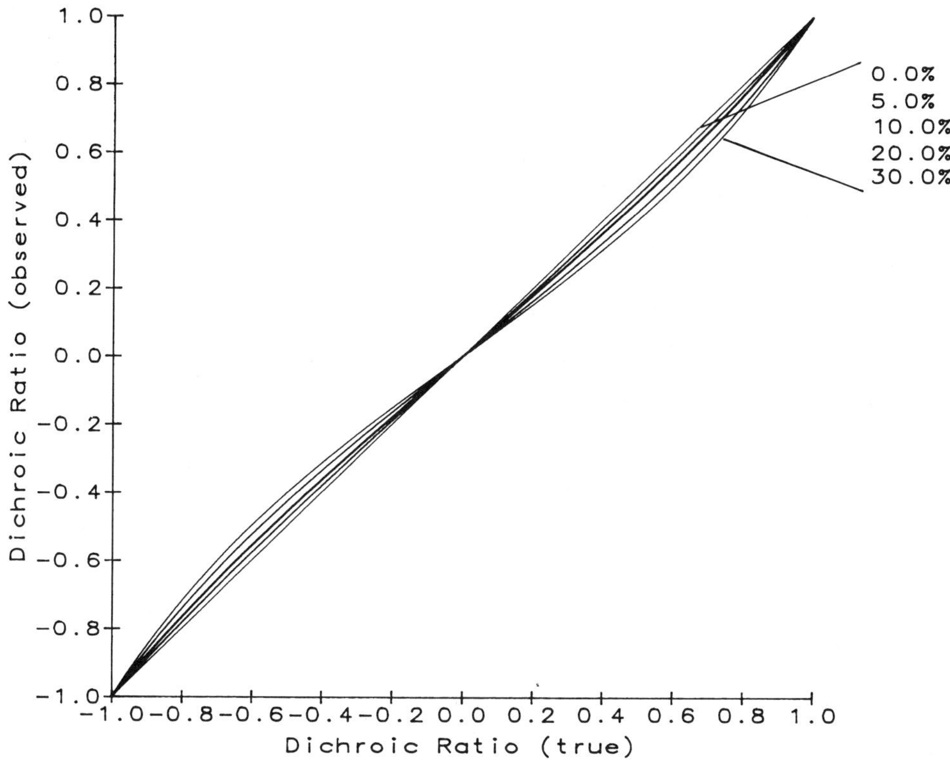

Figure 5. Dichroic ratio error for a band of O.D.=1.0 at varying stray light levels.

K' is an absorption coefficient and the stray light level, ε, can vary between 0 and 1 (0%T and 100%T). It is well known from previous work how the observed absorbance deviates from the true absorbance in the presence of stray light. Figure 4 shows a typical curve of absorbance error for a band with O.D. = 1.0 at varying levels of stray light. The relative error becomes greater as the band becomes more strongly absorbing, i.e. at larger values of K'. For a band with a true absorbance of 1.0, a stray light level of 0.30 will result in absorbance errors of greater than 50%.

It might be expected that the dichroic ratio will show similar behavior. The dichroic ratio is a dimensionless quantity which varies from 1 to -1. This represents a variation in ß, the angle between the fiber axis and the dipole moment vector, of 0 degrees to 90 degrees. If we now vary ß and calculate the measured absorbances in both parallel and perpendicular orientations to the fiber axis with and without stray light, we can plot the results as shown in

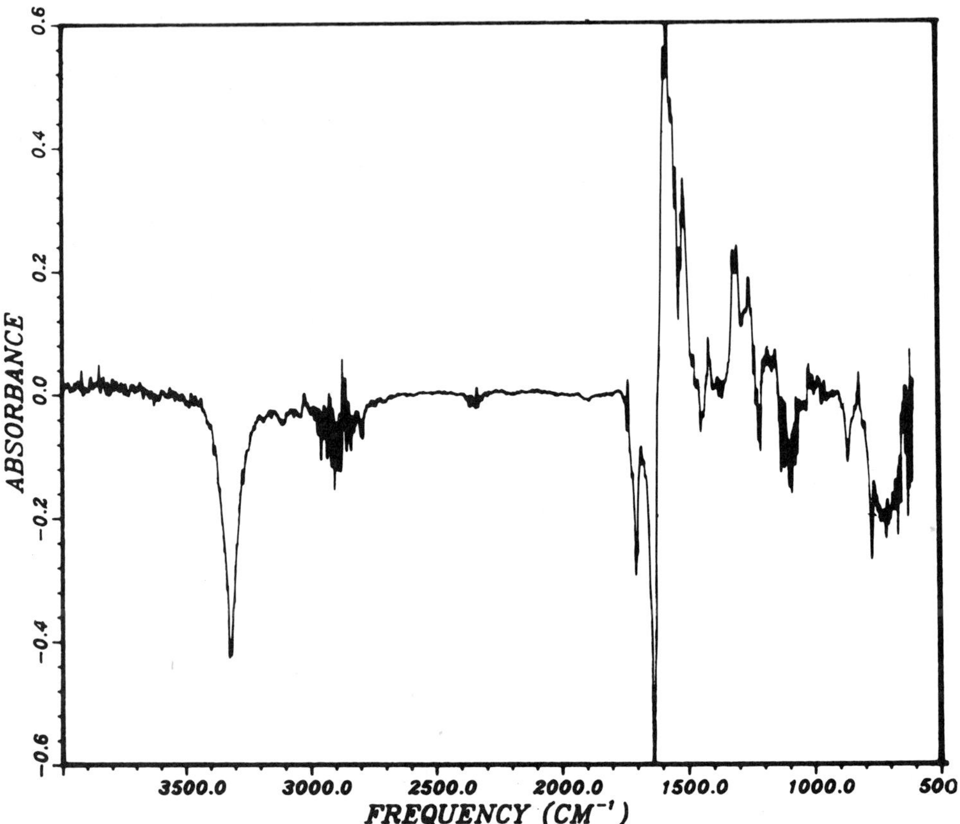

Figure 6. Partial dichroic spectrum of a urethane urea polymer fiber obtained with redundant aperturing.

Figure 5. These are dichroic ratio (D.R.) error curves of observed D.R. versus true D.R for a band of absorbance 1.0 and varying stray light levels. Only at very high values of stray light is there significant deviation, i.e. greater than 2%, from the true dichroic ratio. If we were to plot the same data for a weaker absorption band, O.D.=0.1, the error curves would be indistinguishable from the straight line representing no error. These results involve the assumption that the contribution to stray light is the same in both polarizations, though slight differences in stray light level between polarizations do not affect the results. There are three points where the dichroic ratio is completely independent of the stray light level, and they correspond to values of ß of 0, 45, and 90 degrees. Near these points, the stray light induced error is minimal.

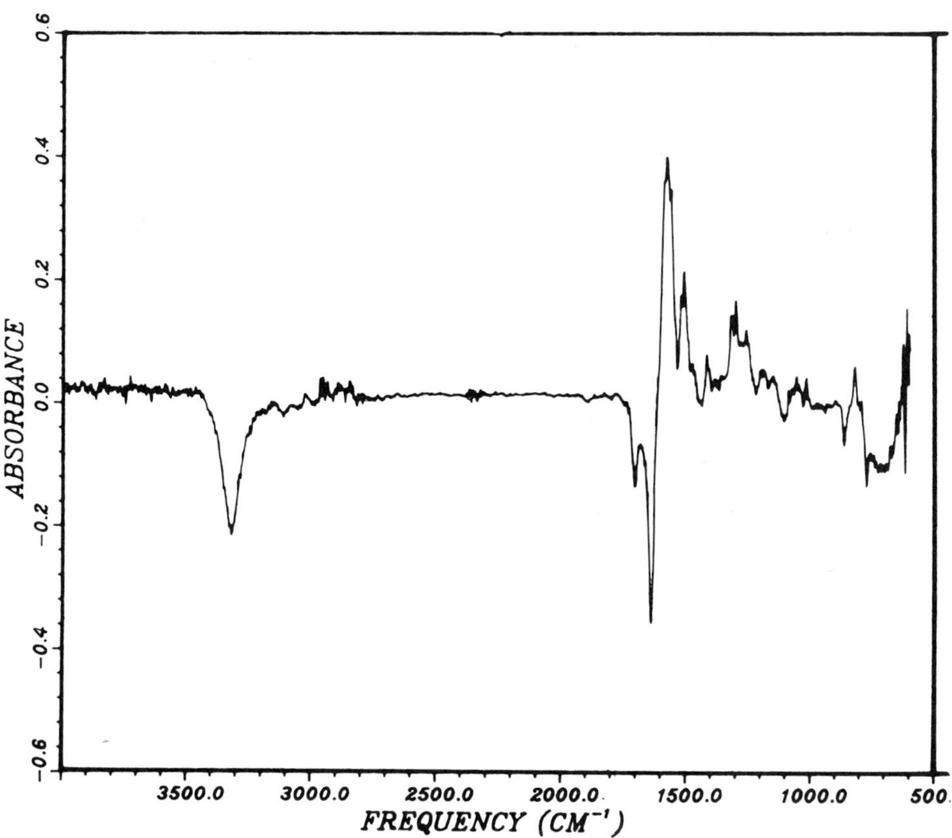

Figure 7. Partial dichroic spectrum of a urethane urea polymer fiber obtained with a single aperture.

If these simulations provide a true picture of reality, we should be able to measure dichroic ratios from spectra obtained with differing stray light levels and find agreement. The optical design of the IR-plan system provides an easy mechanism to make this measurement. A spectrum of a sample can be measured with Redundant Aperturing and then remeasured with a single aperture. Nothing will change except the stray light level due to diffraction effects. Figure 6 shows the partial dichroic spectrum of a urethane urea copolymer fiber which orients under stress. This spectrum was obtained with redundant aperturing. The bands at 1709 cm^{-1} and 1680 cm^{-1} show that the C=O groups orient perpendicularly to the fiber axis as does the N-H group. When one aperture was removed, the partial dichroic spectrum shown in Figure 7 was obtained. The features are the same, but the intensities have changed due to the effects of stray light. Table 1 shows the calculated dichroic ratios for several bands in the spectrum. The agreement between single and redundant aperturing is excellent. This corroborates the observations from the simulations and also helps to explain why earlier work on orientation in fibers succeeded in spite of poor sampling methods. When working with fiber bundles excessive stray light was always present. The observed absorbances were in error, but when the dichroic ratios were calculated, the stray light effect was partially compensated for, and reasonable orientation information was obtained.

Table 1. Calculated dichroic ratios for several bands in the spectra of a urethane urea polymer fiber.

	3300 cm^{-1}	1709 cm^{-1}	1640 cm^{-1}
Single Aperture	-.32	-.14	-.36
Redundant Aperture	-.30	-.12	-.33

7.4 CONCLUSIONS

The incorporation of redundant aperturing into the infrared microscope definitely improves the effective stray light level and allows a much better photometric accuracy. For infrared dichroism studies, if a single aperture system must be used, quantitative orientation information can be obtained only by calculating the full dichroic ratio for the bands of interest. This appears to partially compensate for the effect of stray light contributions from diffraction effects.

8

Applications of Polarized Infrared Microspectroscopy on Polymers

J.W. BRASCH AND A. LUSTIGER* *Battelle Columbus Division, Columbus, Ohio*

8.1 INTRODUCTION

Slow crack brittle-type fracture is the primary mode of material related failure in polyethylene (PE) gas distribution piping. On a molecular level, it is posited that this type of failure occurs through separation of the crystalline regions, or lamellae, in this material under stress [1,2]. Tie molecules are the links which hold these lamellae together. The greater the concentration of tie molecules, the better a given PE will resist slow-crack-type failure. There is, however, no method at present to measure tie molecule concentration.

The process of ductile deformation of a partially crystalline polymer is schematically depicted in Figures 1 and 2. The crystalline portion of a polymer such as polyethylene consists of spherulitic lamellae, represented in these illustrations by the rectangular blocks. The amorphous portion of the polymer consists of loops (molecules with both ends embedded in the same lamella), cilia (molecules with one end embedded in a lamella and the other end free), and tie molecules (molecules whose ends are embedded in adjacent lamellae). Elongation resulting from deformation prior to the yield point involves changes in the amorphous region, i.e., the spherulitic lamellae remain essentially intact. Just after yielding, a fiber morphology is produced, but with essentially the same number of tie molecules. In the amorphous region, the loops are little affected by stretching, and cilia are more likely to retain natural coils and random orientations. The tie molecules, however, are oriented to a very high degree parallel to the direction of stretch.

* *Present Address, AT&T Bell Laboratories, Murray Hill, New Jersey*

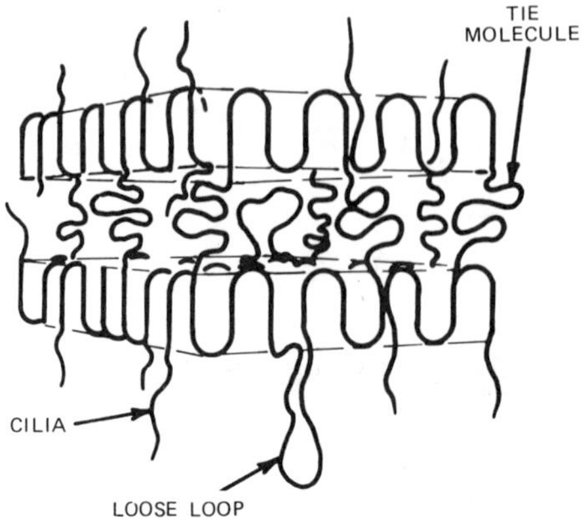

Figure 1. Representation of crystalline and amorphous regions of polyethylene before strain.

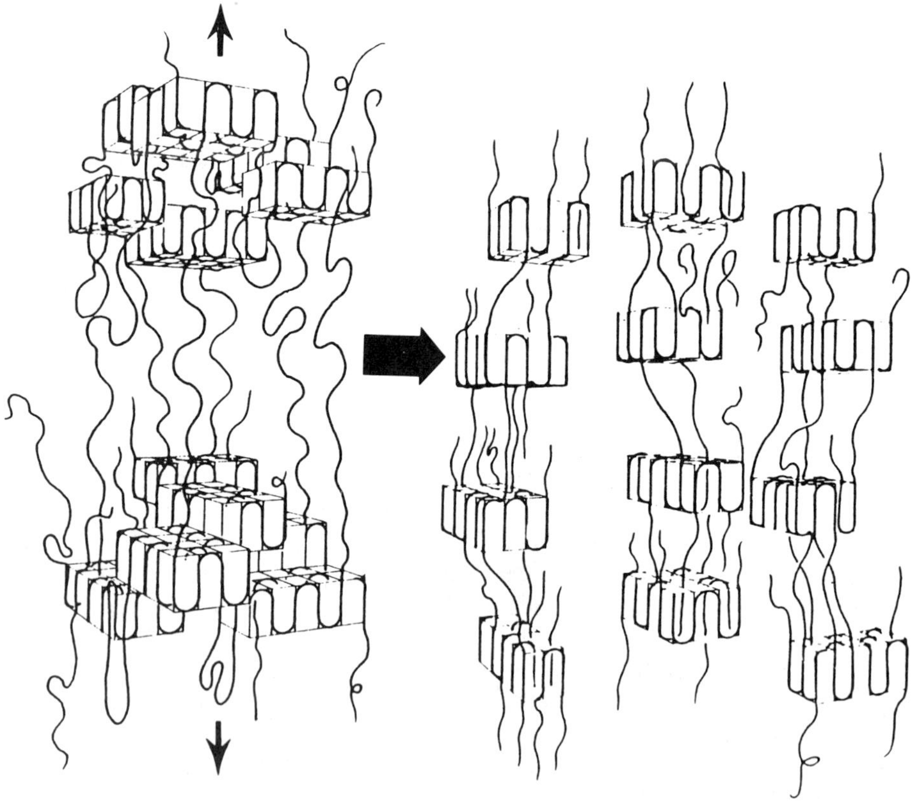

Figure 2. Representation of crystalline and amorphous regions of polyethylene after strain.

This significant increase in parallel orientation of the tie molecules suggests that dichroic infrared spectroscopy would provide a measurement of tie molecule concentration. We therefore conducted a series of experiments on polyethylene. Samples were studied of three widely different molecular weights, but with the same percentage of crystallinity (and conversely the same percentage of amorphous material). The rationale was that if the three samples were subjected to the same elongation, the amorphous region of the low molecular weight sample would consist of a high proportion of cilia and loops and relatively few oriented tie molecules, whereas the high molecular weight sample would have a higher proportion of oriented tie molecules and relatively fewer loops and cilia. Measurable spectral differences might therefore be related to relative tie molecule concentrations.

8.2 EXPERIMENTAL

The polyethylene samples were manufactured by Mitsui Petrochemical (New York, N.Y.), and had the following properties:

Material	Density, g/cc	Melt Index	GPC Molecular Weight Data M_n	M_w	M_z
45,300	0.940	29.0 (low mw)	16,300	54,000	382,200
4,330	0.940	2.4 (med mw)	25,400	102,000	380,000
5,100	0.940	0.23 (high mw)	16,700	222,400	1,630,000

For the initial infrared experiments, 0.2 mm thick films were made by compression molding at 160° C between aluminum sheets. Specimens 13 X 25 mm were cut from these films, clamped in a simple, constant strain fixture and strained 100 percent of their initial length. In each case this was beyond the yield point and produced a neck. The fixture with a strained sample in place is shown in Figure 3. The strain fixture was placed on the stage of an infrared microscope designed and interfaced at Battelle to a Digilab (Cambridge, MA) Model 14 Fourier transform infrared spectrophotometer system. A rectangular slit was imaged onto the polyethylene film. Infrared spectra were obtained for an area of approximately 100 X 500 microns. After numerous measurements, we chose the region approximately 200 microns into the neck for subsequent measurements.

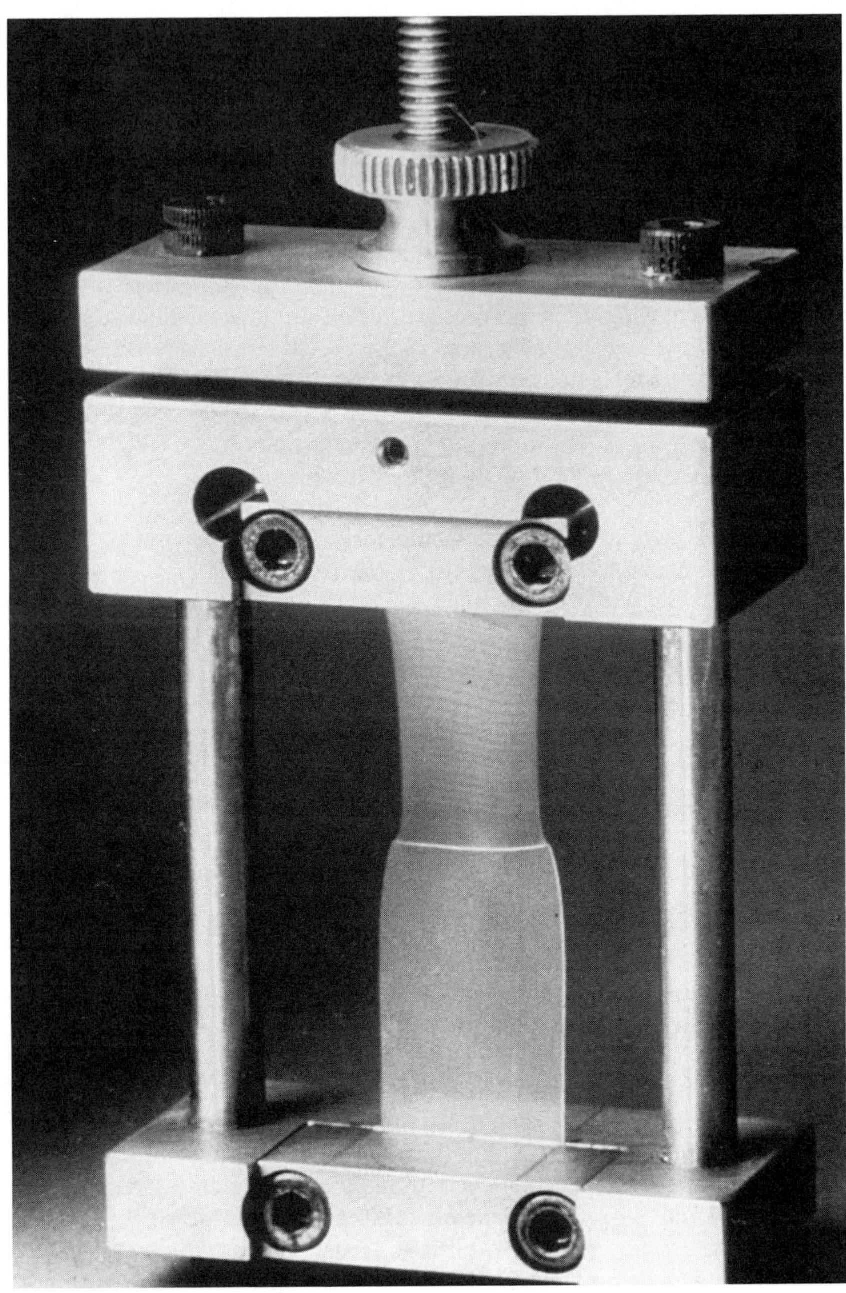

Figure 3. Polymer specimen after 100% strain in the strain fixture.

A wire grid polarizer (Molectron Corp, Campbell, CA) was used for the polarization measurements, oriented parallel and perpendicular to the strain direction. Spectra were obtained at 4 cm^{-1} resolution by coadding 800 scans using a wide band MCT detector. These measurements revealed many differences in dichroic behavior between the three materials, and a few infrared bands which appeared to correlate with molecular weight, but none that could be unambiguously related to tie molecule content. Since the crystalline lamellae can tilt as a result of the applied stress, it is difficult to separate effects of amorphous or crystalline orientation.

A refinement of the experiment was done based on the known fact that halogenation of polyethylene occurs far more rapidly in the amorphous region than in the crystalline region [3, 4]. In essence, halogen atoms could function as a tag on the amorphous material. A second series of experiments were performed which were very encouraging, leading to the final experiment described here.

Films of the three polymers were produced as described before. Specimens for strain were cut from these films except they were only 5-10 mm wide. Single specimens from each of the three polymers were simultaneously clamped in the strain fixture so that all three specimens were treated and measured under exactly equivalent conditions. The fixture was than adjusted to strain 60% of the original length. At this point, polarized and unpolarized spectral measurements were made on the three specimens. The intact fixture with the strained samples was then placed in a container and subjected to vapor phase chlorination (90/10 chlorine/nitrogen) for 1 hour. The spectral measurements were then repeated.

8.3 RESULTS

A typical unpolarized absorbance spectrum of one of the chlorinated polymer samples is shown in Figure 4. The C-Cl stretching vibration ~600 cm^{-1} is discernible as a weak band as expected since the chlorination occurred only near the surface in the amorphous region of these relatively highly crystalline samples.

Figure 5 shows the expanded, unpolarized 700 cm^{-1} region of the three different specimens. The peak height of the 603 cm^{-1} C-Cl stretch, relative to the peak height of the CH_2 rocking mode at 720 cm^{-1}, is approximately the same in all three samples. This indicates that the same level of chlorination occurred in each sample.

Figures 6, 7, and 8 show the expanded, polarized, 700 cm^{-1} region of the

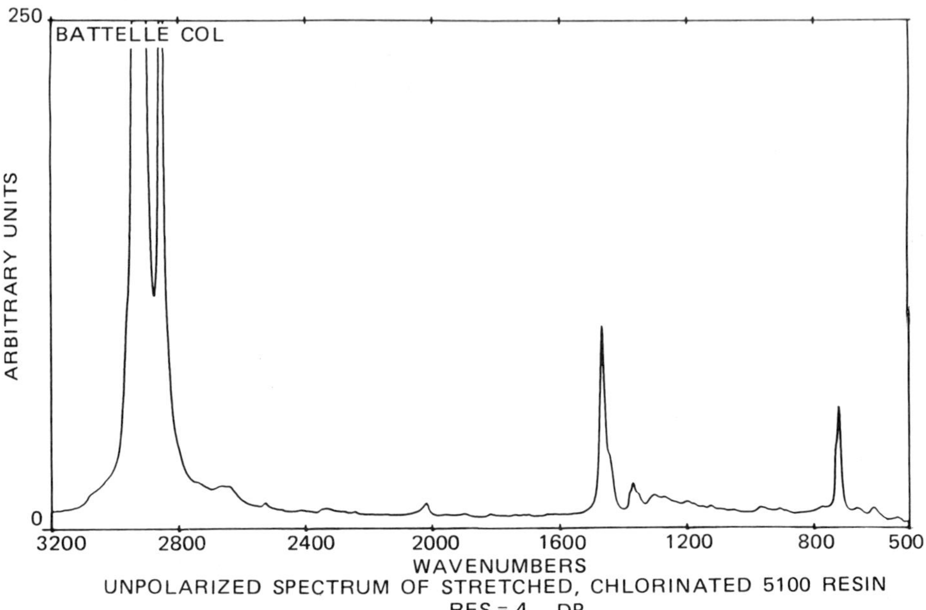

Figure 4. Unpolarized spectrum of a polyethylene specimen after 100% strain and vapor phase chlorination.

low, medium, and high molecular weight polymers, respectively. There is an obvious, progressive increase in dichroism of the C-Cl stretch near 600 cm^{-1} in the three specimens, with the greatest dichroism in the high molecular weight polymer. This trend is made more obvious in Figure 9, where spectral subtractions of perpendicular minus parallel polarized spectra are shown for the three samples. There is almost complete cancellation of the 603 cm^{-1} absorption in the low molecular weight polymer; an obvious residual 603 cm^{-1} absorption in the medium molecular weight polymer; and a strong 603 cm^{-1} band in the high molecular weight polymer (enhanced greatly by a strong interference fringe, but very obviously on top of the last fringe).

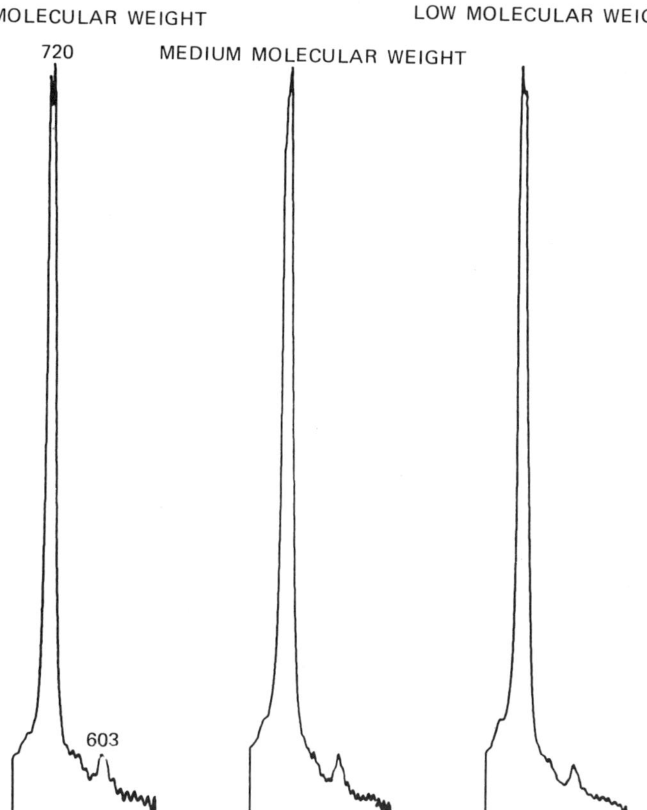

Figure 5. Unpolarized spectra showing ratio of C-C1 peak at 603 cm^{-1} to C-CH_2 peak at 720 cm^{-1}, for high, medium, and low molecular weight polyethylene. Since the ratios are virtually identical, the relative degree of chlorination is equivalent for the three samples.

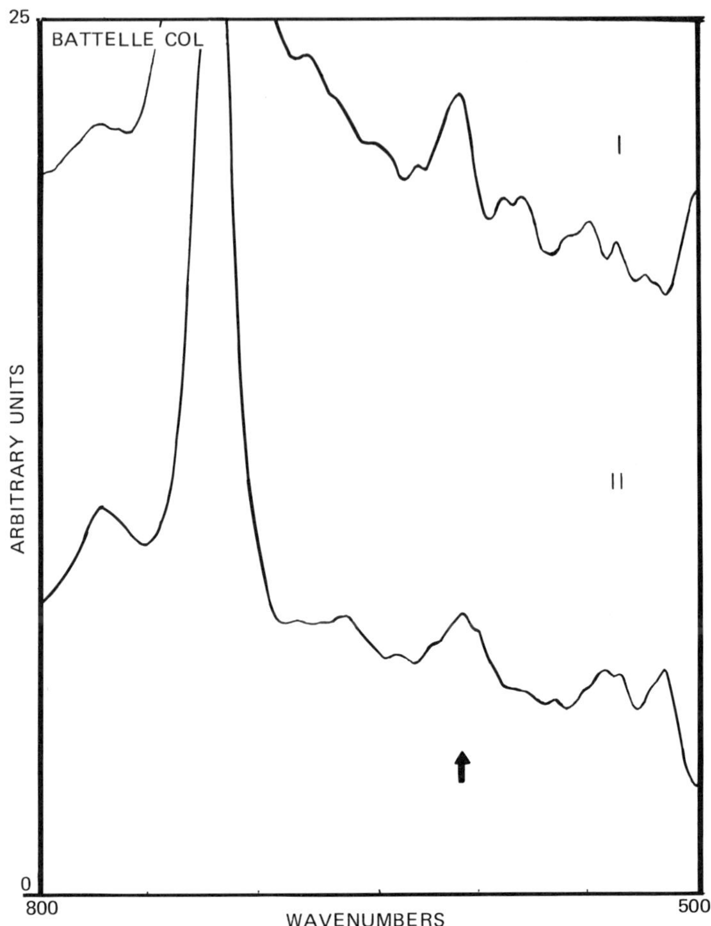

Figure 6. Infrared dichroism of chlorinated 45,300 resin.

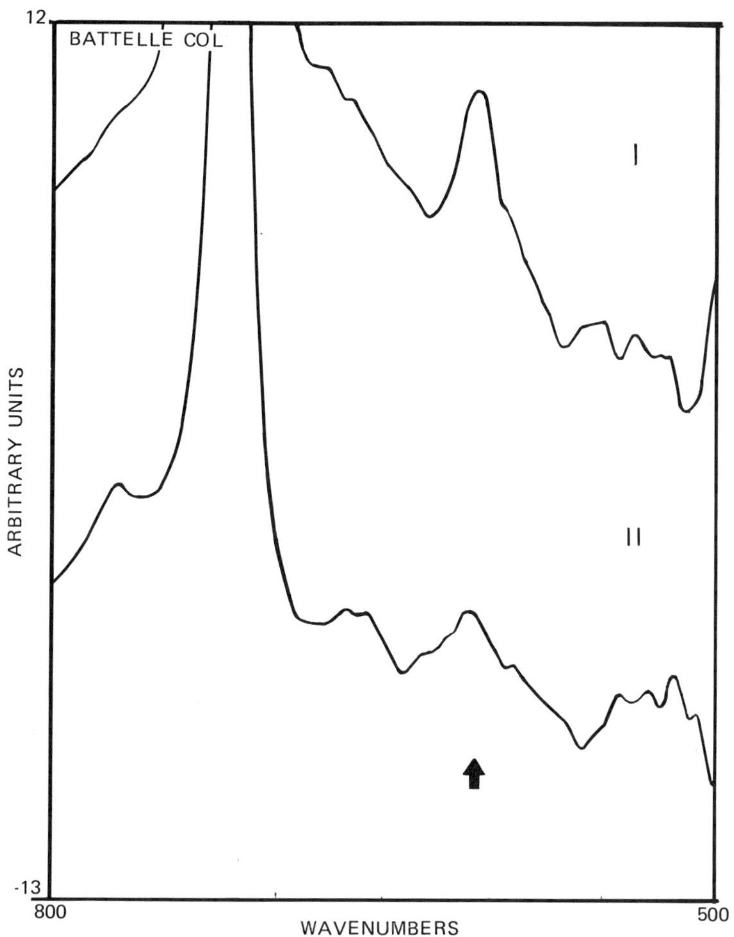

Figure 7. Infrared dichroism of chlorinated 4330 resin.

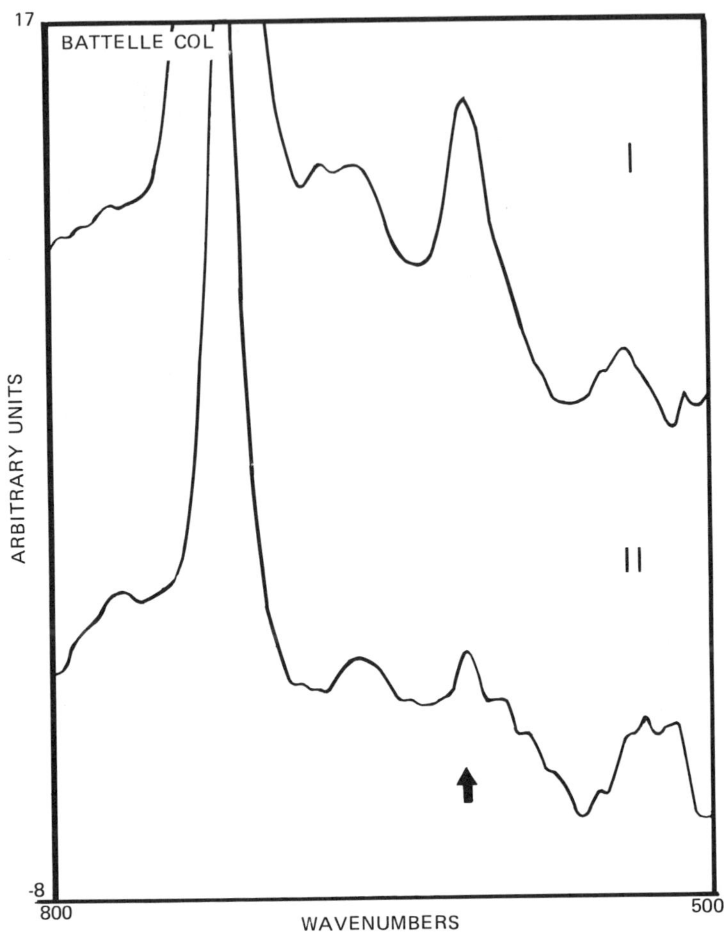

Figure 8. Infrared dichroism of chlorinated 5100 resin.

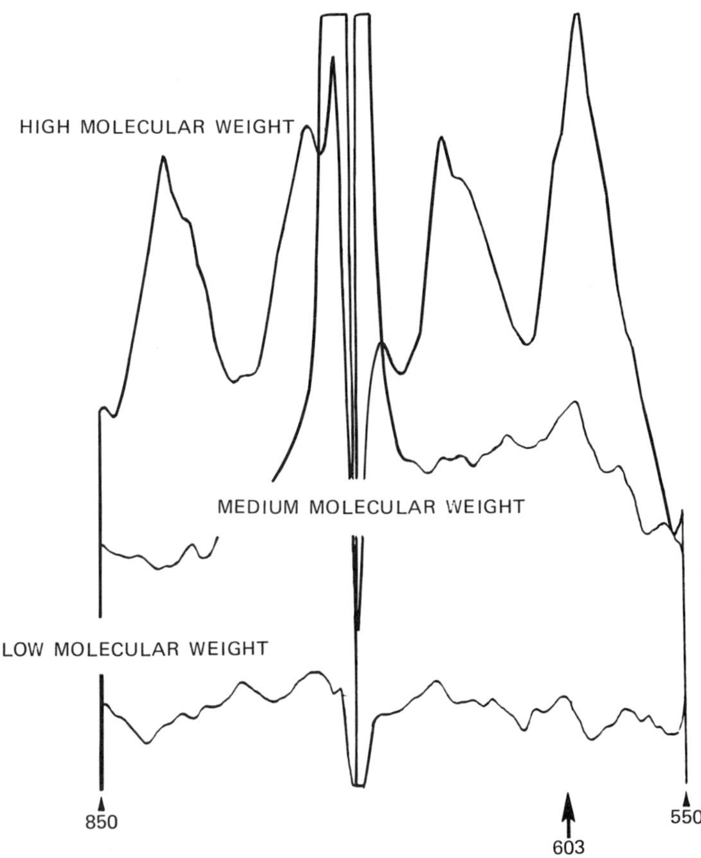

Figure 9. Results of spectral subtraction of polarized spectra of strained, chlorinated polyethylene specimens. In each case, the parallel polarized spectrum was subtracted from the perpendicular polarized spectrum.

8.4 CONCLUSIONS

These data demonstrate that infrared dichroism measurements of the C-Cl stretching vibration of chlorinated specimens are directly relatable to the relative amounts of tie molecules in polyethylene. The ability to determine that parameter will have very important ramifications in understanding and predicting deformation and failure mechanisms in such polymer systems. Use of the microscope in these measurements confines the data to areas of the samples which undergo the greatest change, and therefore improves the sensitivity of the measurement.

ACKNOWLEDGEMENTS

This work was performed at Battelle under Contract 5082-260-0613 funded by the Basic Research Division, Gas Research Institute, Chicago, Illinois.

The authors would also like to express deep appreciation to Robert J. Jakobsen, who presented the paper at FACCS in St. Louis on very short notice.

REFERENCES

1. A. Lustiger, and R. L. Markham, *Polymer*, 24: 1647 (1983).
2. M. J. Hannon, *J. Appl. Poly. Sci.*, 18: 3761 (1974).
3. A. Keller, W. Matreyek, and F. Winslow, *J. Poly Sci.*, 1: 291 (1962).
4. I. Harrison, and E. Baer, *J. Poly. Sci.*, Pt. A2, 9: 1305 (1971).

9

Some Applications of the Polarized FT-IR Microsampling Technique

S. L. HILL AND K. KRISHNAN *Bio-Rad, Digilab Division, Cambridge, Massachusetts*

9.1 INTRODUCTION

The speed and sensitivity advantages of the FT-IR spectroscopy have led to a variety of applications for the methodology. The range of sampling accessories available for use with FT-IR instruments now make it possible to routinely obtain the infrared spectra of practically any chemical compound. Even though it was recognized many years ago [1] that FT-IR spectroscopy can be a very powerful microsampling tool, the technique did not get wide acceptance due to the difficulties associated with the conventional beam condensor-based microsampling method.

However, since the introduction by Bio-Rad Digilab Division (Cambridge, MA) in 1983, of the infrared microscope with all-reflecting optics as the FT-IR microsampling accessory of choice, the FT-IR microsampling method has found a very wide range of applications. Microtransmission and microreflection spectra of a variety of chemical systems such as polymers (multi-layer polymers, and imperfections in polymers), single natural and synthetic fibers (with typical diameters ranging from 10 to 30 micrometers), pharmaceuticals, minerals, coatings on metals (such as Teflon® on razor blades), and semiconductor materials have already been published in the literature [2-7]. The application of the FT-IR microsampling technique to solve practical industrial problems has also been reported [8]. Even the feasibility of obtaining the dichroic transmission spectra from a single poly (ethylene terephthalate) fiber was demonstrated [2]. In this paper such dichroic measurements on a variety of systems will be illustrated. Furthermore, the extension of the microsampling technique into the near infrared spectral region will also be described.

9.2 EXPERIMENTAL DETAILS

The principle of the infrared microscope as the FT-IR microsampling accessory has been described previously [2]. The accessory uses an all-reflecting infrared microscope including a visible illumination system for the visual examination of the samples, and a dedicated, on-axis, small area MCT detector. The small area detector reduces the detector-induced noise, thus leading to higher sensitivity; the sampling size can vary from less than 10 x 10 micrometers in linear dimensions to the dimensions of the detector element used in the accessory. The sample under study is placed on a standard microscope X-Y stage, and can be visually examined under a variety of magnifications. The sample area of interest is isolated by placing a variable aperture on an intermediate image of the sample within the barrel of the microscope. The whole accessory sits atop the FT-IR instrument, and the infrared radiation from the instrument can be transmitted through or reflected from the sample. For the dichroic studies, an infrared transmitting polarizer (gold wire grid supported on KRS-5) is inserted in the optical train above the eyepiece of the microscope as shown in Figure 1. With this arrangement, the sample under study can be visually examined in the usual fashion, the sample area of interest chosen using the variable aperture, and the dichroic measurements made without disturbing the optical train.

For the mid-infrared experiments a narrow range (4,000 - 700 cm^{-1}) MCT detector with a 250 x 250 micrometer element was used. For the near infrared measurements (10,000 - 3,000 cm^{-1}), a 250 x 250 micrometer InSb detector was used. All the spectra reported in this paper were recorded using a Digilab FTS-40 FT-IR instrument. The instrument was equipped with a water-cooled ceramic source, and a Ge on KBr beamsplitter for the mid-infrared measurements. The instrument was fitted with a quartz-halogen source, and an Fe_2O_3 on quartz beamsplitter for the near-infrared experiments. An optical velocity of 1.28 cm/sec was employed, and the spectra were recorded at a spectral resolution of 8 or 4 cm^{-1}. Measurement times of one to two minutes were used to produce the reported spectra. The polarized infrared spectra were recorded using the following procedure: two background spectra were recorded

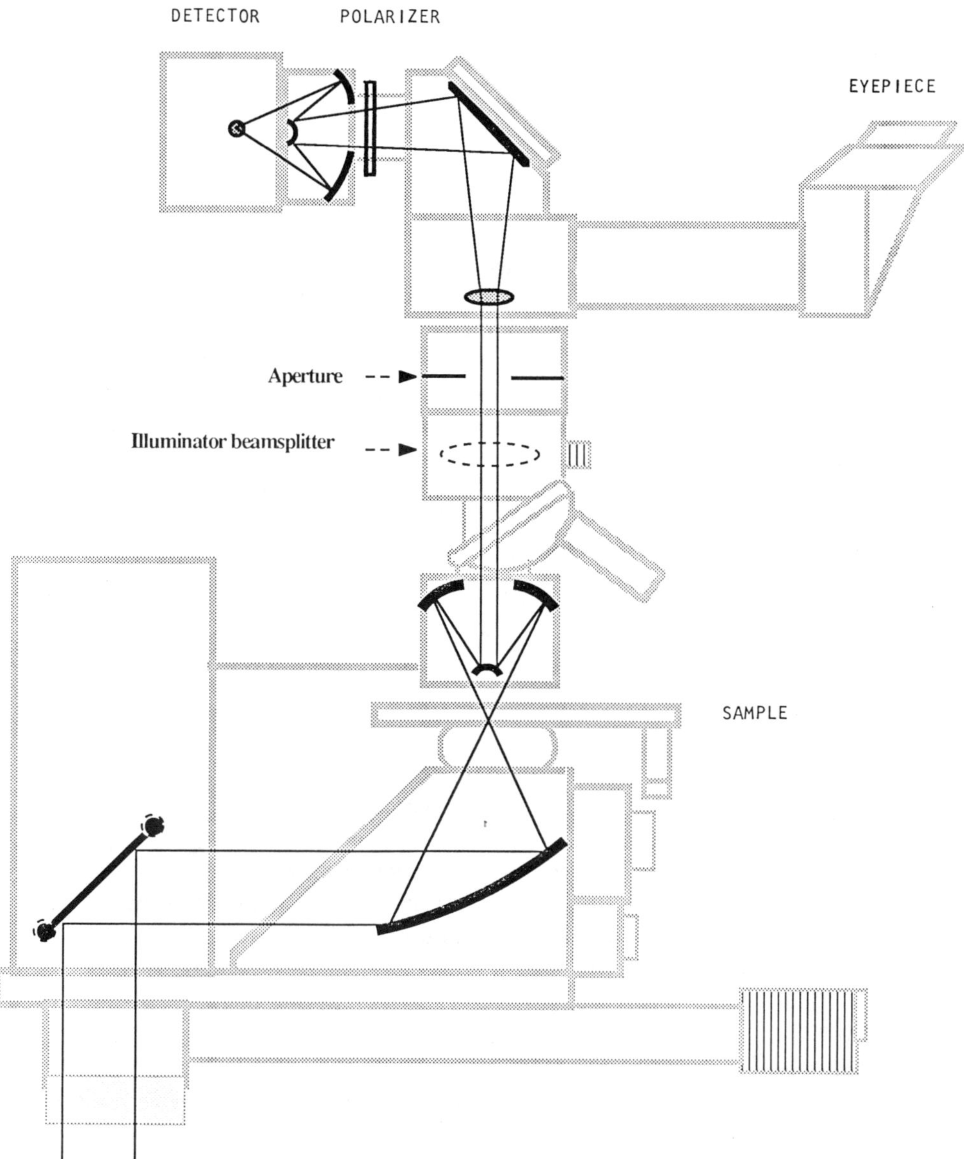

Figure 1. Schematic optical diagram of the Digilab microsampling accessory, including the location of the polarizer.

for the two orthogonal orientations of the polarizer. Then, two sample spectra were recorded for the same two orientations of the polarizer. The dichroic spectra were then created by ratioing the sample spectra against the background spectra for the same polarizer orientation.

9.3 RESULTS AND DISCUSSION

9.3.1 Sensitivity

The sensitivity of the microsampling technique in transmission studies is illustrated by Figure 2. This figure shows the spectum of a polystyrene film with the aperture set to sample a 5 x 5 micrometer area. The background necessary to produce this transmission spectrum was also recorded with the aperture set to the same dimensions. One can observe the high signal-to-noise ratio of the spectrum, particularly down to 1000 cm^{-1}. Below 1000 cm^{-1} one observes some spectral distortions due to diffraction effects. One would normally expect to see diffraction effects dominate the whole spectrum, as the sampling size used in this case is comparable to the wavelength of the infrared radiation. However, most of these diffraction effects are balanced out when the sample and the background spectra are both recorded using the same aperture setting.

9.3.2 Dichroic Measurements on Single Crystals

There is considerable interest in performing dichroic measurements on single organic or inorganic crystals. These experiments are somewhat difficult on a macroscopic scale since fairly large single crystals will have to be grown, and then cut polished down to a few micrometers thickness for the infrared transmission measurements. With the microsampling technique, the sample preparation becomes very easy, and one can use a single crystallite, a few micrometers in linear dimensions, for performing the measurements. This is illustrated in Figure 3 which shows the dichroic spectra of a single crystallite of pyrelene ($C_{20}H_{12}$), recorded with the sampling area set to 100 x 120 micrometers. This planar ring compound exists in the form of yellow-colored platelets. Under the microscope, a single platelet of thickness around 10 micrometers can be selected, and oriented on a KBr transmission plate, and the dichroic measurements made on the sample. The two dichroic spectra recorded from such a sample are shown in Fig. 3. The two spectra are plotted on the same absorbance scale, and one can see from Fig. 3 that whereas the ring modes show little or no dichroism, the modes associated with the carbon-hydrogen

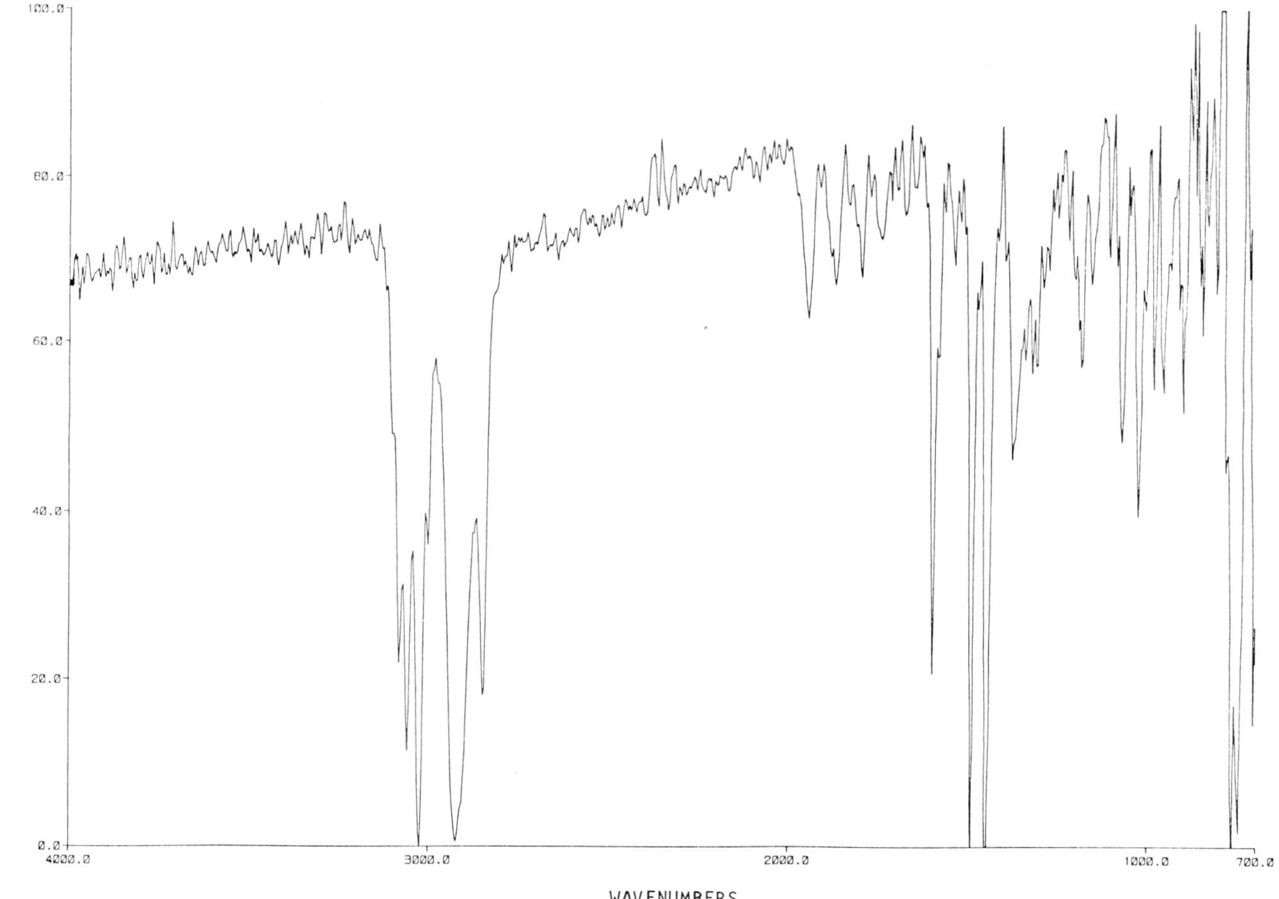

Figure 2. Transmission spectrum of a polystyrene film with sampling area set to 5x5 micrometers. The spectrum was recorded at 8 cm^{-1} resolution, coadding 1,000 scans.

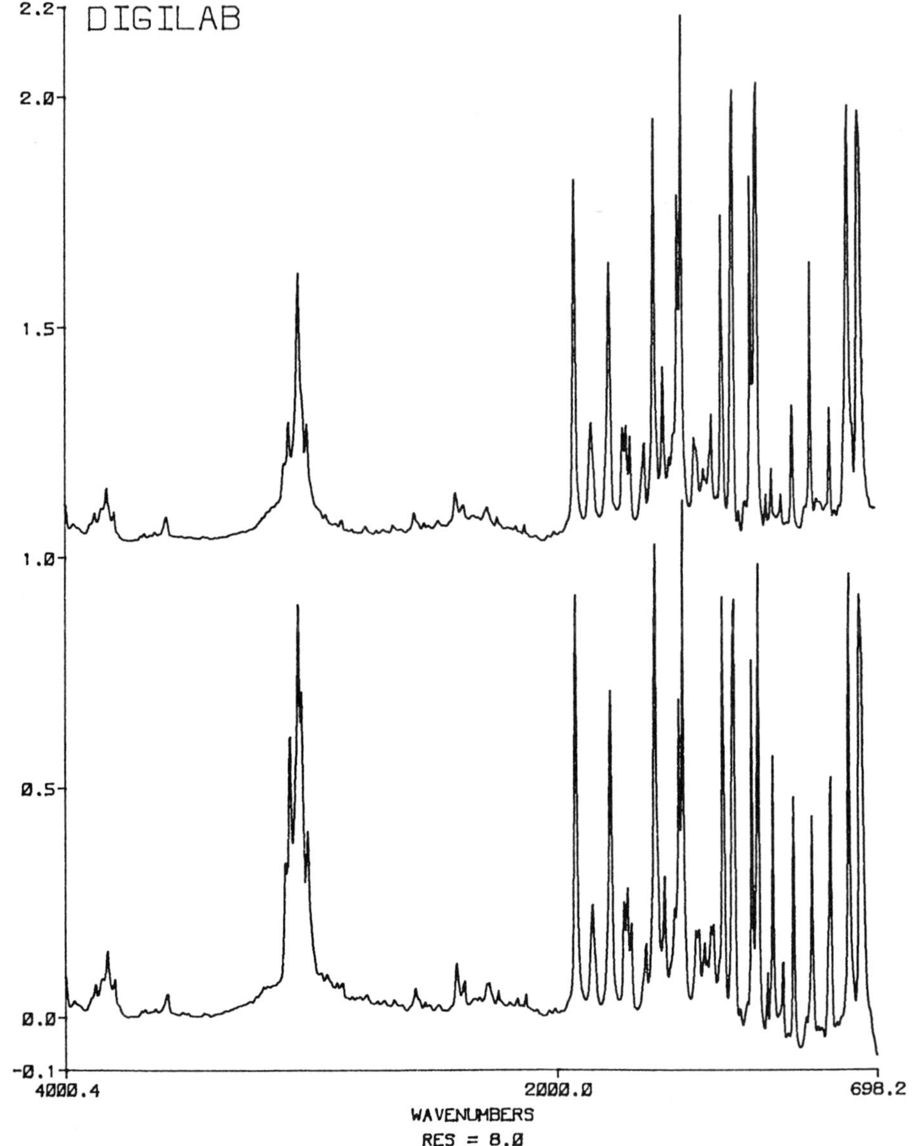

Figure 3. Dichroic spectra of a single crystal of pyrelene ($C_{20}H_{12}$). The spectra were recorded at 8 cm^{-1} resolution, coadding 512 scans, with the electric vector oriented parallel to (bottom) and perpendicular to (top) the long axis of the crystal.

vibrations exhibit varying degrees of polarizations. As one would expect, the C-H out-of-plane modes between 700 and 1,000 cm^{-1} exhibit the most dichroic behaviour. The example shown in Fig. 3 illustrates the ease with which such difficult experiments could be carried out.

Another example of dichroic measurements on a single crystallite is shown in Figure 4. The sample in this case was nickel phthalocyanine doped with BF_4 [9]. These rod-shaped crystallites will be electrically conducting at high doping levels, and will not exhibit any infrared transmission. However, the particular crystallite used in this study (through the courtesy of Prof. C. R. Kannewurf, Northwestern University, Chicago, IL) was inadequately doped, and transmitted the infrared radiation. The dichroic spectra of this sample were recorded with the sampling area set to 120 x 40 micrometers, and the polarizer electric vector orientations set parallel and perpendicular to the long crystal axis. The two dichroic spectra are plotted in transmission on the same scale. One can see that the dipole moments in this crystal are very highly oriented to such an extent that one can hardly see any spectral features in one of the dichroic spectra.

9. 3. 3. Dichroic Measurements on Oriented Polymers

Orientation measurements in polymers using vibrational spectroscopy is a topic of considerable interest, and a number of papers have appeared in the literature on this subject (see Reference 10 for a review). Using the microsampling technique it is now possible to study the polymer orientations on a microscopic level. As mentioned in the introduction, the feasibility of performing dichroic measurements on a single fiber poly (ethylene terephthalate, 15 micrometer diameter has already been described elsewhere (2).

Figure 5 shows the dichroic spectra of a 30 micrometer thick, unstretched poly (butylene terepthalate) film. The spectra were recorded by coadding 64 scans each at 4 cm-1 resolution, with the sampling area set to 200 x 200 micrometers. As one can see from the figure, the spectra exhibit little or no dichroism. The sample was then subjected to a 3x uniaxial draw, and the dichroic spectra were recorded as before. These spectra are shown in Figure 6. The lower spectrum was obtained with the electric vector oriented along the draw axis. One can clearly see the dichroic behaviour of the drawn polymer. The dichroism exhibited by this sample can be explained in terms of the gauche to trans transformation on stretching. A more detailed study of this polymer system is underway, with particular emphasis on the characterization of its dichroic behaviour in the “necking” region of the stretched polymer. A detailed analysis of these results will be published elsewhere.

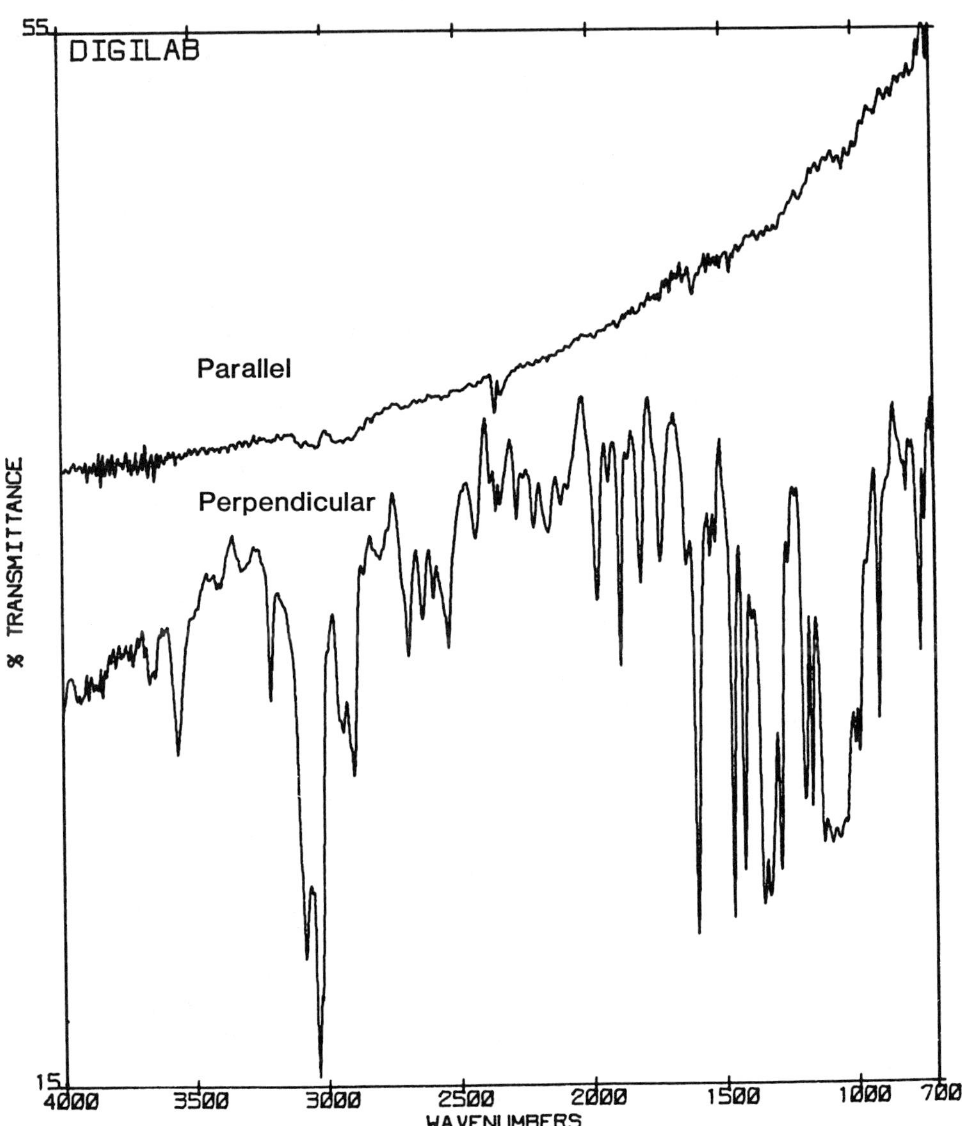

Figure 4. Dichroic spectra of a single crystal of nickel phthalocyanine doped with BF_4. Experimental conditions were the same as for Fig. 3.

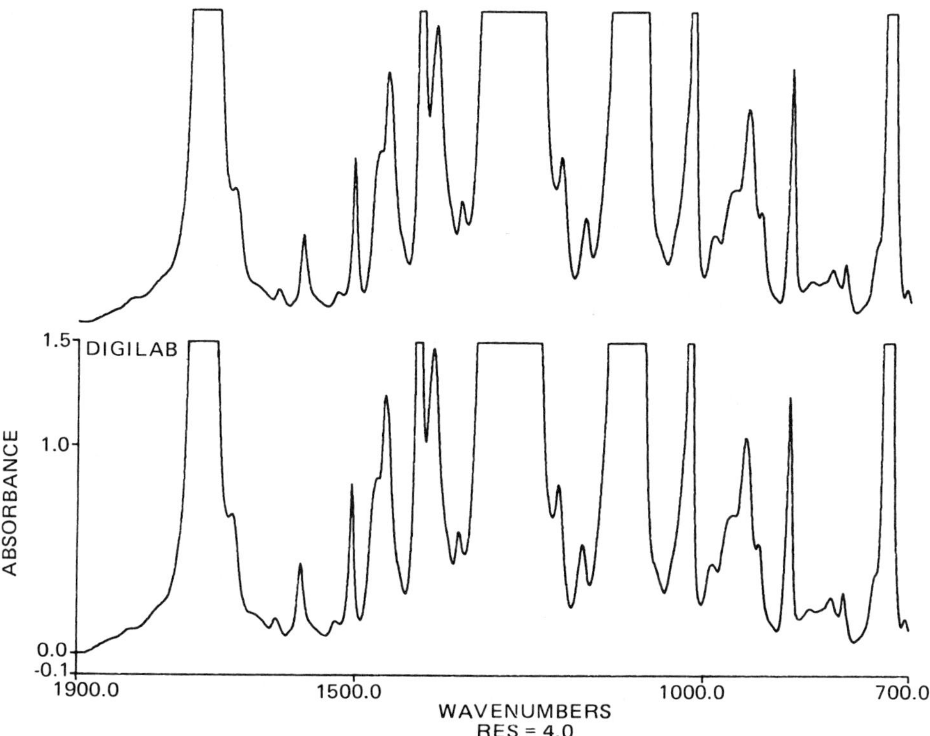

Figure 5. Dichroic spectra of unstretched poly (butylene terephthalate), PBT, film. The spectra were recorded at 4 cm^{-1} resolution, coadding 64 scans.

9. 3. 4. Microsampling in the Near Infrared

The microsampling accessory can very easily be adapted for use in the near infrared region by replacing the MCT detector by an InSb detector. Overtone and combination bands, and electronic transitions can be studied by using this method. When one wants to study the transmission spectra in the mid-infrared, the sample thickness must be cut down to the 10 - 20 micrometer range, usually by microtoming the sample (8). Much thicker samples can be used in the near infrared. Figure 7 shows the near-IR microtransmission spectra of each layer in a two layer polymer sample. The sample in this case was a commercial plastic

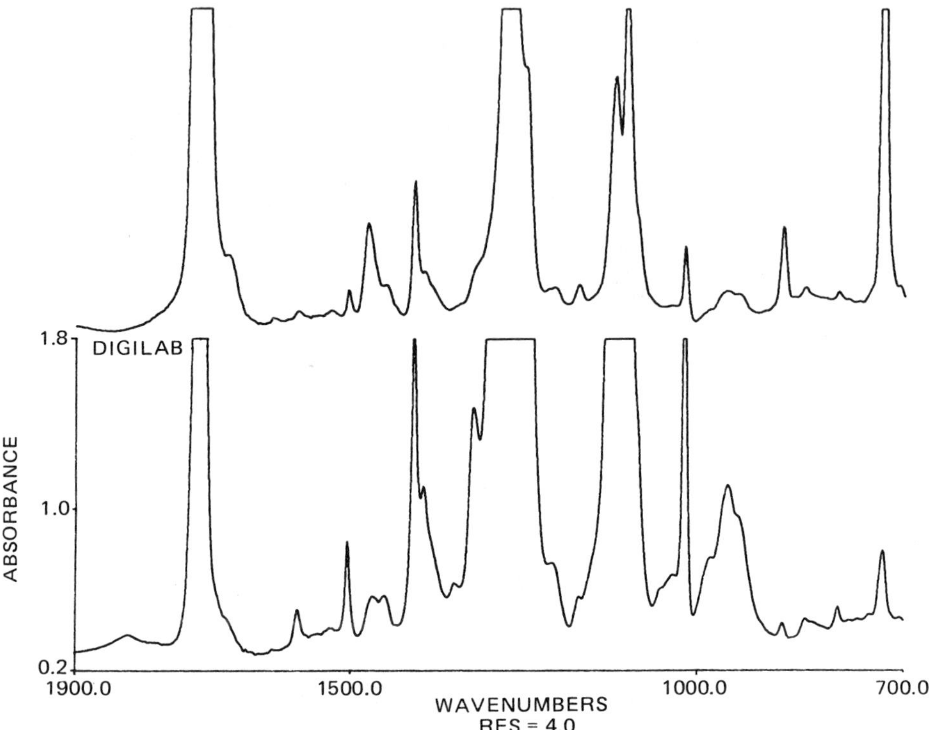

Figure 6. Dichroic spectra of 3x-drawn PBT film. The electric vector was oriented parallel (bottom), and perpendicular (top) to the draw axis.

container, and an approximately 50 micrometer thick section of this sample was cut using a sharp razor blade. The two polymer layers that made up this sample were clearly visible, and their spectra were recorded with the sampling area set to 50 x 100 micrometers. One can see the overtone and combination bands in the 4,000 - 10,000 cm^{-1} region in the figure. The lower spectrum in the figure is that of nylon, and the upper spectrum is that of polypropylene.

Another example of a near-IR microtransmission spectrum is shown in Figure 8. The sample used in this case was a tiny (80 x 120 micrometer) crystallite of $LaCuCl_4$ kept in a nitrogen atmosphere inside a sealed fused quartz capillary tube. The spectrum of this sample was recorded through the capillary

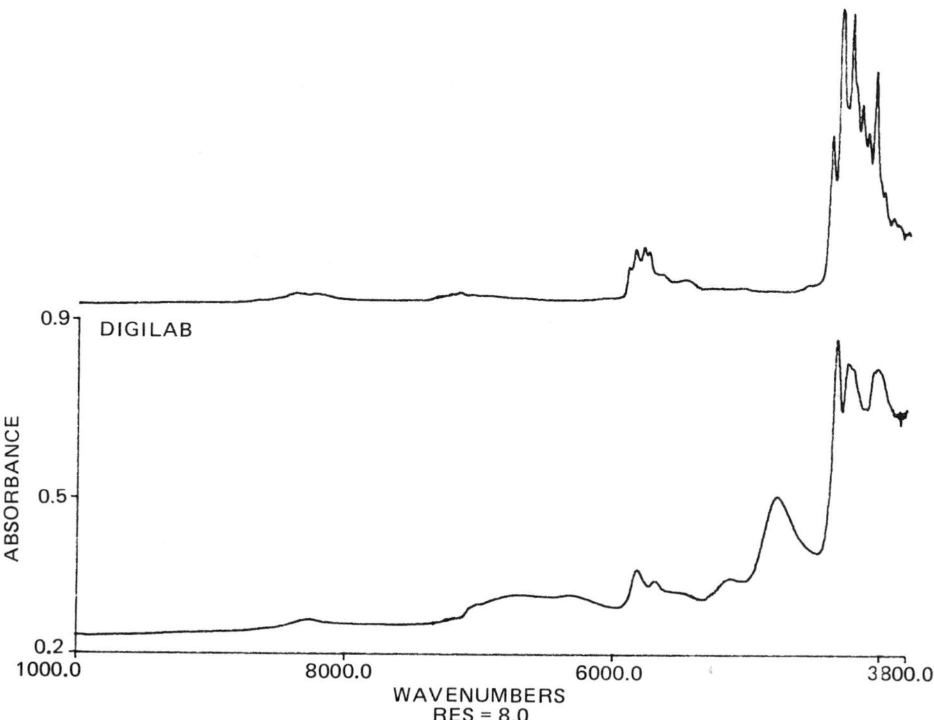

Figure 7. Near infrared transmission spectra of each layer of a two-layer commercial polymer laminate - nylon (bottom), and polypropylene (top). These spectra were recorded at 8 cm^{-1} resolution, coadding 128 scans.

tube. Some of the bands seen in the spectrum are due to overtone and combination bands, and the others are due to low-energy electronic transitions. A detailed paper dealing with the assignments of the electronic transitions in this and similar rare-earth inorganic compounds will be published elsewhere.

An example of dichroic measurements in the near infrared using the microsampling technique is shown in Figure 9. The sample used in this case was a relatively thick (ca. 25 micrometer) single crystallite of chlorobenzene. One can clearly see the dichoic behaviour of the C-H stretching bands, as well as the overtone, and combination bands. Such spectra will be useful in the assignement of the various bands, and estimating the molecular orientations in the unit cell of the crystal.

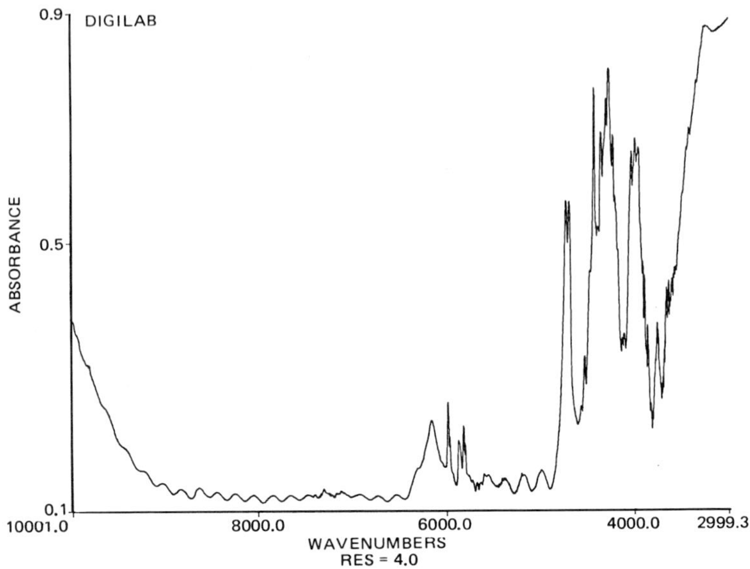

Figure 8. Near infrared transmission spectrum of a single crystal of $LaCuCl_4$. Experimental conditions were the same as for Fig. 7.

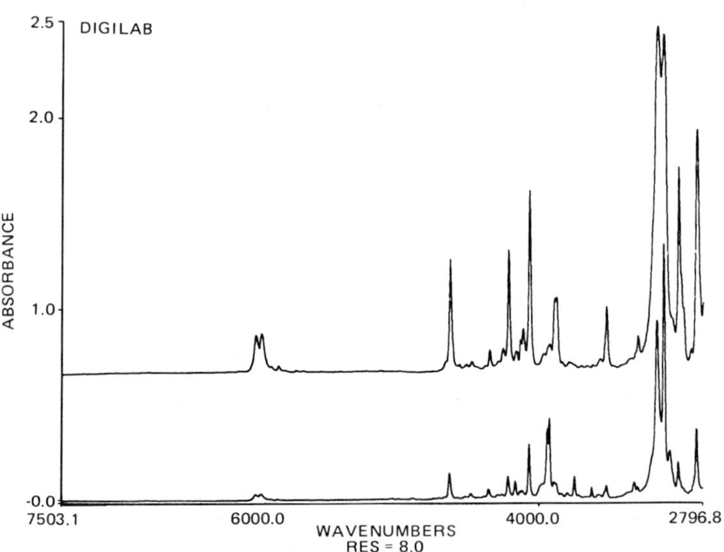

Figure 9. Near infrared dichroic spectra of a single crystal of chlorobenzene. The electric vector was oriented parallel (bottom), and perpendicular (top) to the long crystal axis. 128 scans at 4 cm^{-1} resolution were coadded to produce these spectra.

9. 4 CONCLUSION

It has been shown that the FT-IR microsampling technique can be used for polarization studies on a variety of chemical systems. Extension of the microsampling technique to the near infrared region has also been demonstrated.

REFERENCES

1. R. Cournoyer, F. C. Shearer, and D. H. Anderson, *Anl. Chem.*, 49: 2275 (1979).
2. K. Krishnan, *Polymer Preprints*, 25: 182 (1984).
3. K. Krishnan, S. L. Hill, and L. S. Gelfand, in *International Conference on Fourier and Computerized Infrared Spectroscopy* (J. Grasselli, and D. G. Cameron, eds.), SPIE, Billingham, WA, 665: 252 (1986).
4. A. Ishitani, Ibid, 553: 25 (1985).
5. K. Krishnan, in *Optical Techniques for Industrial Inspection* (P. G. Cielo, ed.), SPIE, Billingham, WA, 665: 252 (1986).
6. K. Krishnan and D. L. Kuehl, in *Semiconductor Processing* (C. G. Gupta, ed.), ASTM STP, Philadelphia, PA, 850: 325 (1984).
7. K. Krishnan and R Mundhe, in *Spectroscopic Techniques in Semiconductor Technology*, SPIE, Billingham, WA, Vol. 452 (1983).
8. M. A. Harthcock, L. A. Lentz, B. L. Davis, and K. Krishnan, *Appl. Spectoc.*, 40: 210 (1986).
9. P. Ainabe, S. Nakamura, W-B. Liang, T. J. Amrks, R. L. Burton, C. R. Kennewurf, and K. Imaeda, *Jour. Amer. Chem. Soc.*, 107: 7229 (1985).
10. B. Jasse and J. L. Koenig, *J. Macromol. Sci. -Rev. Macromol. Chem.*, C17: 61 (1979).

Application of Infrared Microspectroscopy to the Semiconductor Industry

10

Infrared Microspectroscopic Analysis as Applied to Problems in Semiconductor and Packaging Materials and Processes

KAREN MADDEN, BARBARA BERGIN, NANCY KLYMKO, JOHN N. RAMSEY *IBM Corporation, Z/41D, East Fishkill, Hopewell Junction, New York*

10.1 INTRODUCTION

Semiconductor technology continues to drive toward higher chip densities and smaller device dimensions, thus demanding more stringent requirements on process and manufacturing cleanliness as well as creating a need for more sophisticated techniques in small area analysis. Elemental small area/particle analysis has been successfully accomplished by the use of an electron microprobe, scanning electron microscopy with energy dispersive analysis, and scanning Auger electron spectroscopy. Molecular analysis of crystalline materials was accomplished by the use of X-ray diffraction camera techniques or extraction replication and electron diffraction in a transmission electron microscope. With the advent of infrared microspectroscopic analysis, the ability to get a molecular analysis of contamination that previously was beyond the size limitation of conventional infrared has now become possible. The technique of infrared microspectroscopy has been available for use at our facility for the past six years (Nanospec 20/Micro-IR, Nanometrics, Sunnyvale, CA) and although there were limitations (sample size, limited reflectance capability, computer control enhancements) we have gained invaluable experience in preparing microsamples for infrared analysis using a microscope accessory, and have solved many problems. In addition, the complementary small area technique of Raman spectroscopy has been successfully used in conjunction with infrared microspectroscopy. The use of these various techniques has been reviewed recently [1]. Now with the introduction of a microscope attachment for FT-IR, we have been able to qualitatively analyze micro samples that were once beyond the limitations of the instrumentation.

The infrared microspectroscopy techniques have been especially useful in identifying microscopic contamination that occurs during the development and manufacture of semiconductor and packaging components. Such contamination can be the result of process residues, handling and storage contaminants, and/or products of reactions that occur during processing. This paper will describe three cases in which we, an analytical laboratory which supports both site development and manufacturing of computer components, were asked to identify particle contamination and locate the source(s).

10.2 INSTRUMENTAL

The infrared spectra included in this chapter were obtained using an Analect (Irvine, CA) AQS-20 FT-IR with microscope attachment and a Nanospec 20/IR microspectrophotometer. Both instruments were equipped with MCT (Mercury Cadmium Telluride) detectors.

10.3 APPLICATIONS

10.3.1 Semiconductor Technology

The photolithographic process in the manufacturing of semiconductor chips involves numerous processing steps. During one such step, silicon wafers that have been coated with photoresist are placed in drying ovens at 230°C for curing of the photoresist. It was in one of these ovens that, during routine maintenance on the oven, a bright yellow, powdery film was found on parts of the oven. There was insufficient sample to make a KBr pellet for conventional macro-infrared analysis. However, it was possible to transfer an aggregate, approximately 50μm in size, to a KBr disk and to analyze the sample by infrared microspectroscopic analysis.

Suspected to be the source of contamination was a coumarin dye, which is added to the photoresist as an anti-reflective agent to prevent image deformation during photolithographic processing. The physical characteristics of the oven residue (color, consistency) as well as the bands in the infrared spectrum (N-H and C=O stretching at 3339 cm^{-1} and 1696 cm^{-1}, respectively), suggested that the oven residue was due to the sublimation of the dye from the resist and, indeed, this was found to be the case. Figure 1 shows the spectrum obtained for the contaminant and the spectrum of the coumarin dye.

Another type of contamination problem found in the semiconductor processing area involved sliver-like particles that were discovered on silicon wafers after deposition of a polyimide insulating layer. These particles were approximately 1Oµm by 1OOµm and their presence on the wafers resulted in further processing problems. Since silicon is transparent to infrared radiation, we initially tried to analyze the particles *in situ*, however the resulting spectrum was similar to the reference spectrum of polyimide, indicating the particles were coated with polyimide or the molecular structure of the particle was not infrared active. An attempt to remove the particle from the surface of the wafer resulted in splintering the particle, thereby making it too small to be analyzed by this technique, in addition to the possibility of it still being coated with polyimide.

At this point it was observed that certain fresh bottles of the polyimide had tiny slivers floating on top of the liquid. The polyimide is stored in amber glass bottles in refrigerators set at approximately 32°C. The slivers were removed and washed with a solvent to remove any residual polyimide. Figure 2 shows a sliver, free of polyimide, and its infrared spectrum. From this spectrum generalizations could be made about the molecular structure of this specimen which supplied clues to its identification. First, the band located at 3300 cm^{-1} is indicative of N-H stretch. The bands at 1500 cm^{-1} (with weak overtone at approximately 3100 cm^{-1}) and at 1240 cm^{-1} to 1250 cm^{-1} are the result of CNH bending and stretching in the opposite and same direction, respectively. The presence of a carbonyl structure is evident by the band at 1630 cm^{-1}.

Armed with this spectral information and information about the process, we were able to define a reasonable model of what was occurring in the polyimide solution and forming these particles. The particles were identified as dicyclohexyl urea (DCU) which is formed when dicyclocarbodiimide (DCC) (a reagent used in the formation of the polyimide) reacts with water. It is believed that excess DCC was not removed from the product and during a time of high humidity, this excess material reacted with water resulting in the formation of DCU. DCU was synthesized and analyzed by infrared microspectroscopy and was shown to match that of the particles removed from the polyimide. This finding was verified by complementary mass spectrographic analysis.

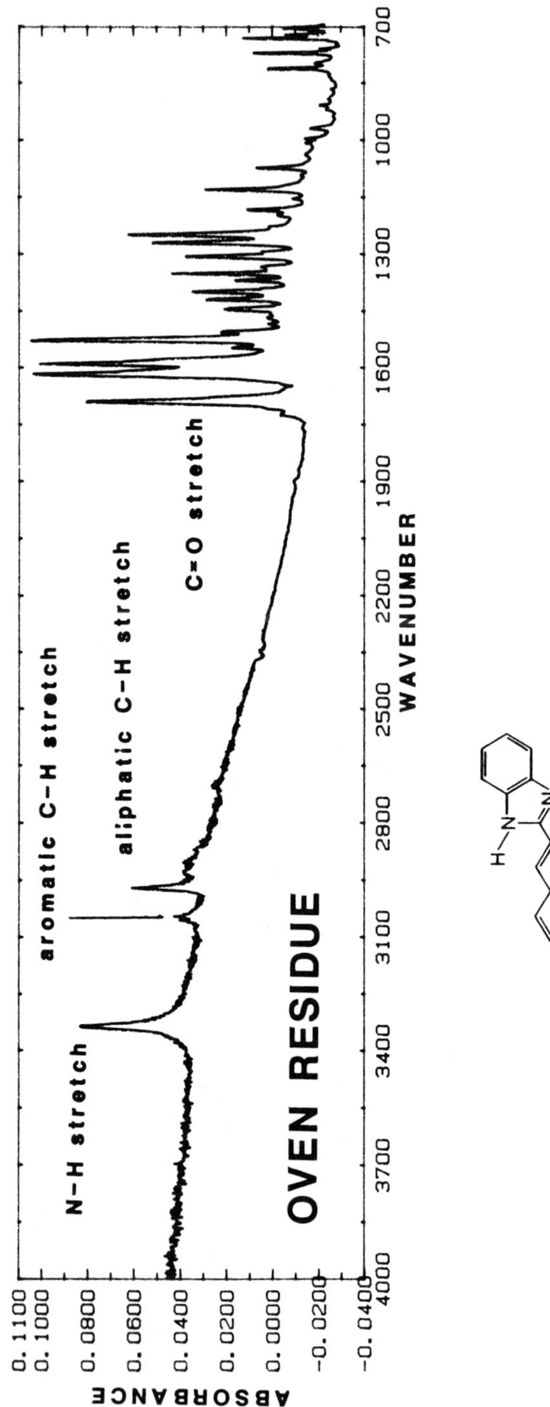
OVEN RESIDUE
N-H stretch
aromatic C-H stretch
aliphatic C-H stretch
C=O stretch
ABSORBANCE
WAVENUMBER
0.1100
0.1000
0.0800
0.0600
0.0400
0.0200
0.0000
-0.0200
-0.0400
4000
3700
3400
3100
2800
2500
2200
1900
1600
1300
1000
700
$(H_5C_2)_2N$
H
N
O

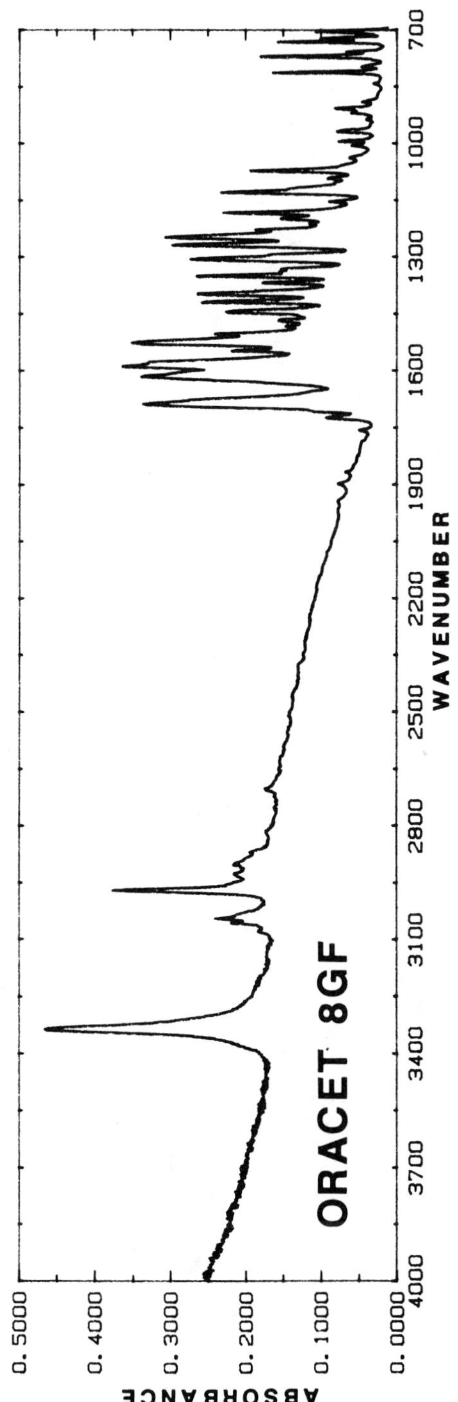

Figure 1. Spectrum a: Oven residue; Spectrum b: Coumarin dye.

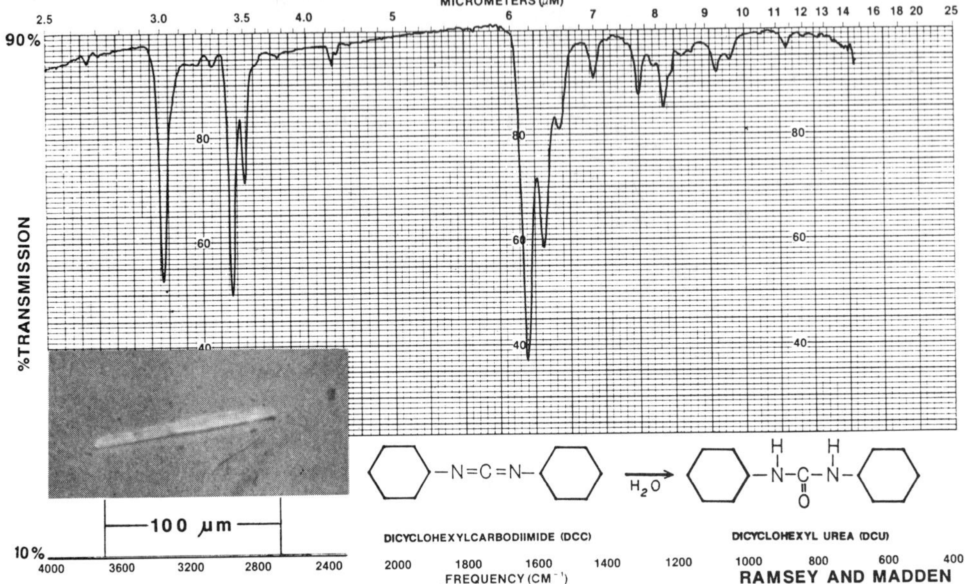

Figure 2. Dicyclohexyl Urea.

10.3.2 Packaging Technology

Thermal paste is used as a heat dissipation medium in packaging technology. This paste mixture must be free of foreign particulate matter. Particles in the paste can result in poor heat dissipation resulting in the overheating of a chip and/or package.

It was in one such package that, when disassembled, two particles were found after inspection of the thermal paste. The particles were removed and carefully washed with hexane to remove the paste. Particle #1, Figure 3, was analyzed and found to be poly (ethylene terepthalate) or polyester. The infrared spectrum is shown in Figure 3. The second particle found in the paste, Figure 4, was identified as being polystyrene. The infrared spectrum of Particle #2 is shown in Figure 4. Although the specific source of these particles would be difficult to locate because of the broad usage of both polystyrene and polyester, we are confident that these particles came from packing materials (Particle #1) and articles of clothing (Particle #2), both of which are common materials

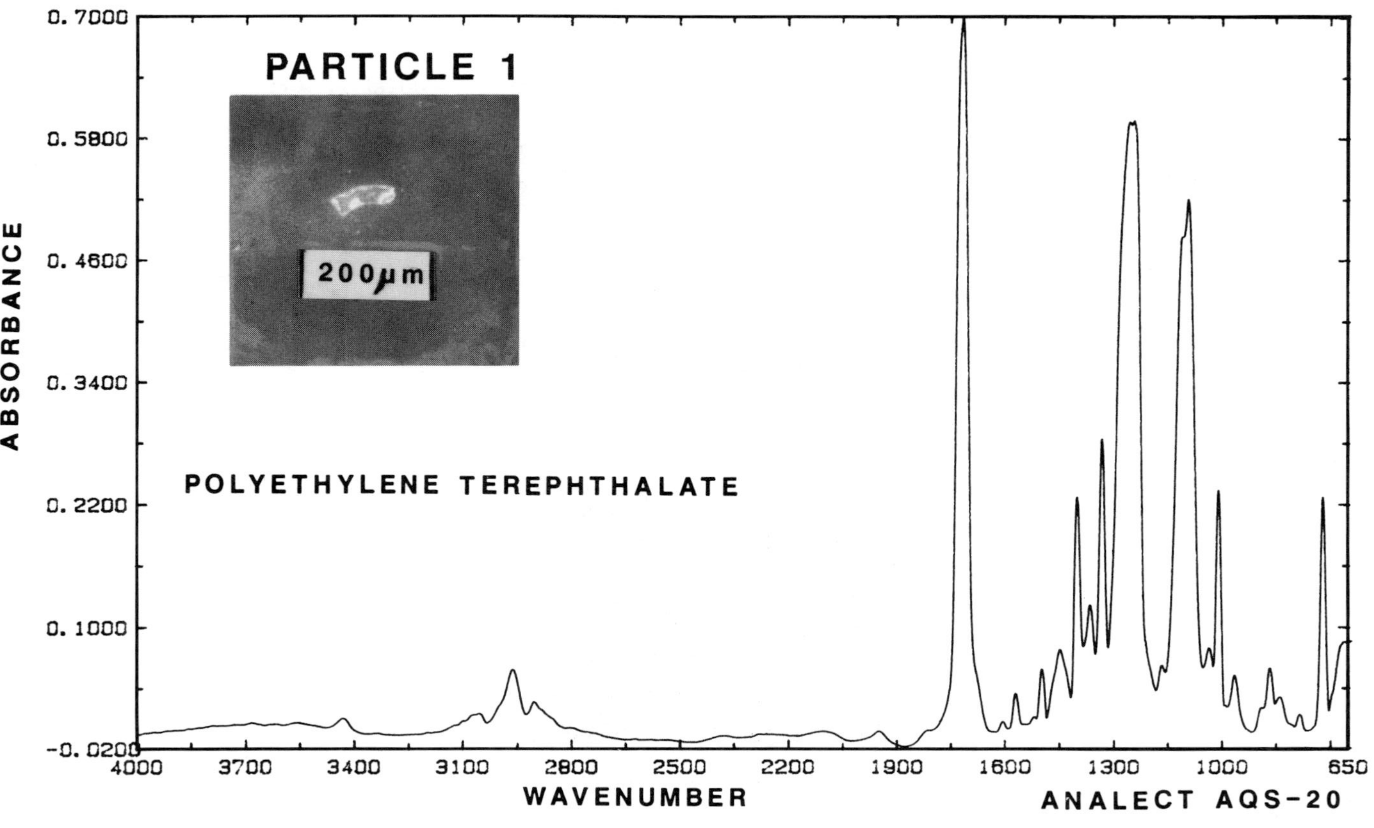

Figure 3. Poly (ethylene Terephthalate).

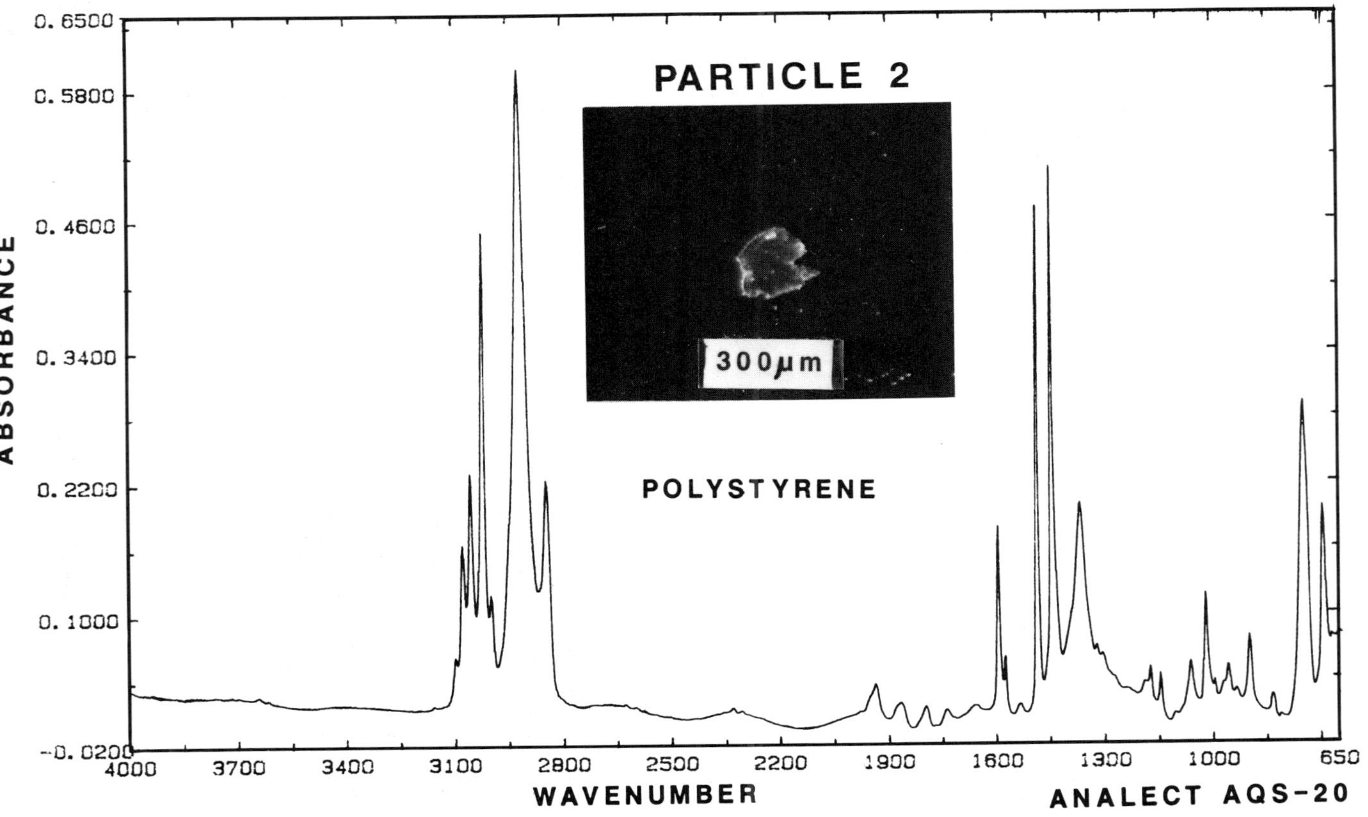

Figure 4. Polystyrene.

found in our environment. This example demonstrates the need for protection from common forms of contaminants when dealing with the manufacture of micro electronic devices and their packages.

10.4 SUMMARY

Infrared microspectroscopy is a powerful technique which has enabled the analytical lab to better support semiconductor/packaging development and manufacturing. With the current commercially available instrumentation it is now possible to detect and identify microscopic contaminants which previously could not be analyzed. The identification of the contaminants is essential in determining the sources of contamination which plague the stringent manufacturing conditions necessary for the production of denser devices and smaller geometries.

REFERENCES

1. Ramsey, J. N., *Microelectronics Processing Problem Solving: The Synergism of Complementary Techniques*, in *Microelectronics Processing, Inorganic Materials Characterization* (L. E. Casper, ed.), ACS Symposium Series #295 (1986).

11

Characterization of Semiconductor Silicon Using the FT-IR Microsampling Techniques

K. KRISHNAN *Bio-Rad, Digilab Division, Cambridge, Massachusetts*

11.1 INTRODUCTION

Infrared spectroscopy is a valuable tool for the characterization of silicon properties during various stages of device manufacturing. Interstitial oxygen and substitutional carbon concentrations in high resistivity silicon can be measured routinely using infrared absorption spectroscopy. Thicknesses of silicon epitaxial layers can be measured from the reflection of the infrared radiation from the silicon sample. The concentrations of phosphorous and boron oxides in the silicate passivation layers on silicon can be determined quantitatively. The concentrations of hydrogen in the silicon nitride passivation layers on silicon can also be measured in a similar fashion. The above measurements are normally carried out over relatively large areas of the sample - typically 3 to 8 mm in diameter. It has been shown [1, 2] that by using Fourier transform infrared microspectroscopic techniques these measurements could be carried out over sample areas as small as 50 x 50 micrometers. In this paper the applications of these FT-IR microsampling techniques to silicon characterization will be presented.

11. 2 EXPERIMENTAL DETAILS

The Digilab universal microsampling accessory, described elsewhere [3,4] which can be used for both the transmission and reflection measurements was used in these experiments. For the epitaxial thickness and the boron and phosphorous concentration measurements, a 250 x 250 micrometer narrow range (4,000 - 700 cm^{-1}) MCT detector was used; for the oxygen and carbon measurements a 250 x 250 micrometer wide range (4,000 - 450 cm^{-1}) MCT detector was used. All measurements were carried out with the Digilab FTS-60 or the FTS-40 FT-IR instruments. For measurements of the oxygen and carbon concentrations and the epitaxial thickness in silicon, a computer-controlled, motorized X-Y stage was used in place of the manual stage on the microsampling accessory.

11. 3 RESULTS AND DISCUSSION

11. 3. 1 Epitaxial Thickness Measurements

Figure 1 shows the schematic of a silicon epitaxial layer on silicon. The epitaxial layer normally ranges in thickness from 0.5 to 200 micrometers depending upon the particular device to be built. The substrate is heavily n- or p-doped, and the epitaxial layer has no doping or is only lightly doped. Lightly doped silicon is transparent to the infrared radiation, and the heavily doped silicon is opaque to most of the infrared radiation due to the free carrier absorption. Thus, when the infrared radiation is made to fall on the epitaxial sample, reflection of the radiation takes place from the surface of the epitaxial layer as well as from the epi-substrate interface. The resulting infrared spectrum will show interference fringes due to the thickness of the epitaxial layer. In the case of FT-IR spectroscopy, these fringes will appear as a spike or secondary interferogram when the interferogram from the FT-IR instrument is reflected from the sample. By measuring the retardation between the main interferogram and the secondary interferogram, the thickness of the epitaxial layer can be calculated precisely. This mechanism of secondary reflection from the epitaxial structure is shown in Figure 1.

Figure 2 shows the single scan interferograms recorded at 8 cm^{-1} resolution from an epitaxial film of nominal thickness around 30 micrometers. The substrate was n-doped with arsenic and the epi was lightly n-doped with phosphorous. The interferograms shown were recorded, from bottom to top, with the aperture set to 100 x 100, 50 x 50, 25 x 25, and 10 x 10 micrometers.

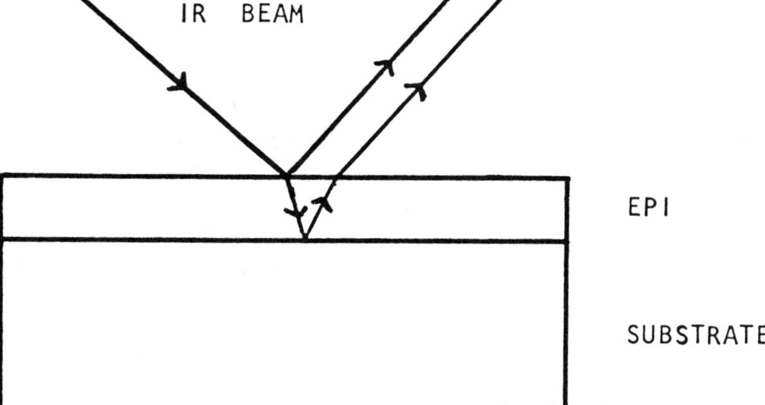

Figure 1. Schematic representation of the reflection of infrared radiation from an epitaxial layer.

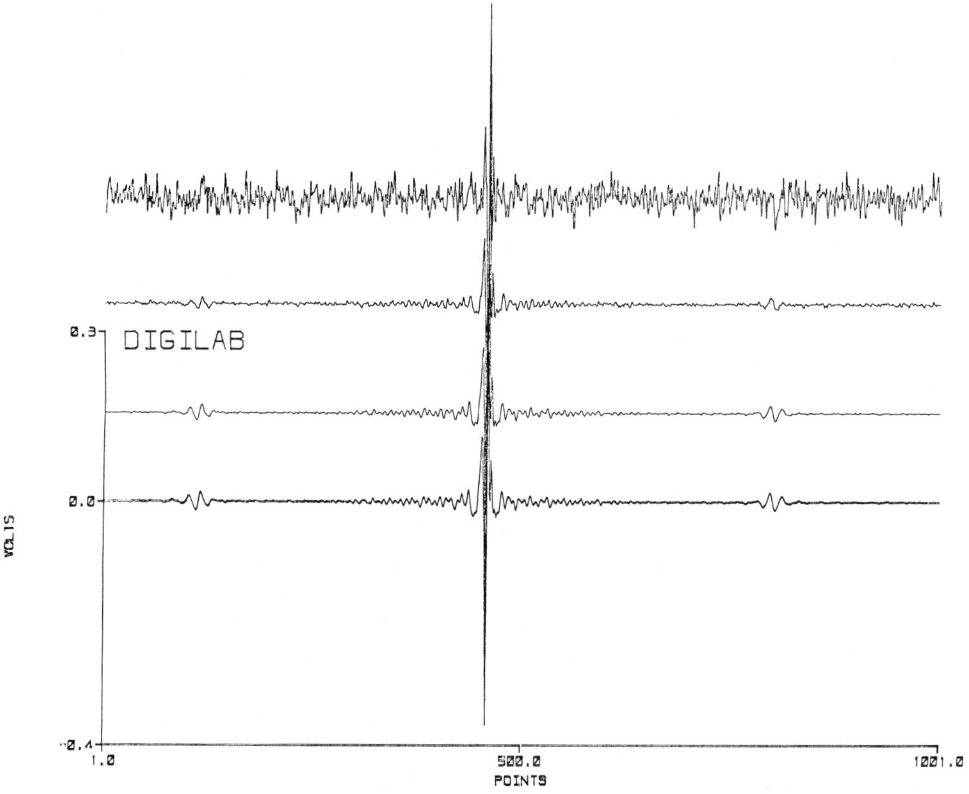

Figure 2. Single-scan interferograms from a 30 micrometer thick silicon epitaxial layer. From bottom to top, the sampling areas were 100x100, 50x50, 25x25, and 10x10 micrometers, respectively.

In the lower three interferograms, one can clearly see the secondary interferogram due to the epitaxial film thickness. The 10 x 10 micrometer single-scan interferogram is too noisy, and the secondary interferogram is not visible above the noise. Figure 3 shows a 256 scans co-added interferogram with the aperture set to sample 10 x 10 micrometer area of the sample. The noise has been reduced considerably, and the secondary interferogram is clearly visible.

In the current semiconductor technology, the density of devices on the silicon wafer is very high, and it is common practice to dope the substrate over very small areas of the samples using ion implantation techniques. The ion implanted areas will usually be the order of a few square micrometers. An epitaxial layer will then be deposited over a whole silicon wafer. A given device wafer may contain many such buried layers, and each buried layer may be doped with a different element. During the actual device manufacturing steps, the silicon sample will be thermally treated, and some out-diffusion of the substrate dopants into the epitaxial layer may occur. The length of this out diffusion will be different for the different dopant elements. This effect is shown schematically in Figure 4. It is very important to know the precise epitaxial thicknesses over each buried layer for the proper functioning of the finished device. An example of the epitaxial measurement over a buried layer structure is shown in Figures 5 and 6.

Figure 5 shows the photograph taken through the microsampling accessory of a silicon wafer with a number of buried layers. These layers are defined by the rectangular features seen in the figure. Any infrared radiation that falls within the rectangular areas will be reflected from the epi-buried layer interface. The resulting interferogram will show the secondary interferogram. Outside the buried layer areas, the silicon substrate is not doped heavily, and these areas will be transparent to the infrared radiation. The reflected interferograms from these areas will not show the secondary interferogram. This latter interferogram can be used as the reference interferogam, which could be subtracted away from the buried layer interferogram to define the secondary interferogram from the buried layer. This subtraction process will be very necessary when the epitaxial film thickness is small (less than 10 micrometers) as in these cases the secondary interferogram will be too close to the main interferogram to be clearly visible before the main interferogram has been subtracted away.

Figure 6 shows the interferograms recorded by co-adding 64 scans at 8 cm^{-1} resolution from a 20 x 20 micrometer buried layer (bottom), from an adjacent area outside this buried layer (middle), and the difference interferogram (top). One can clearly see the secondary interferogram in the top figure. A thickness value of around 5.5 micrometers can be determined for this epitaxial layer from the position of the secondary interferogram.

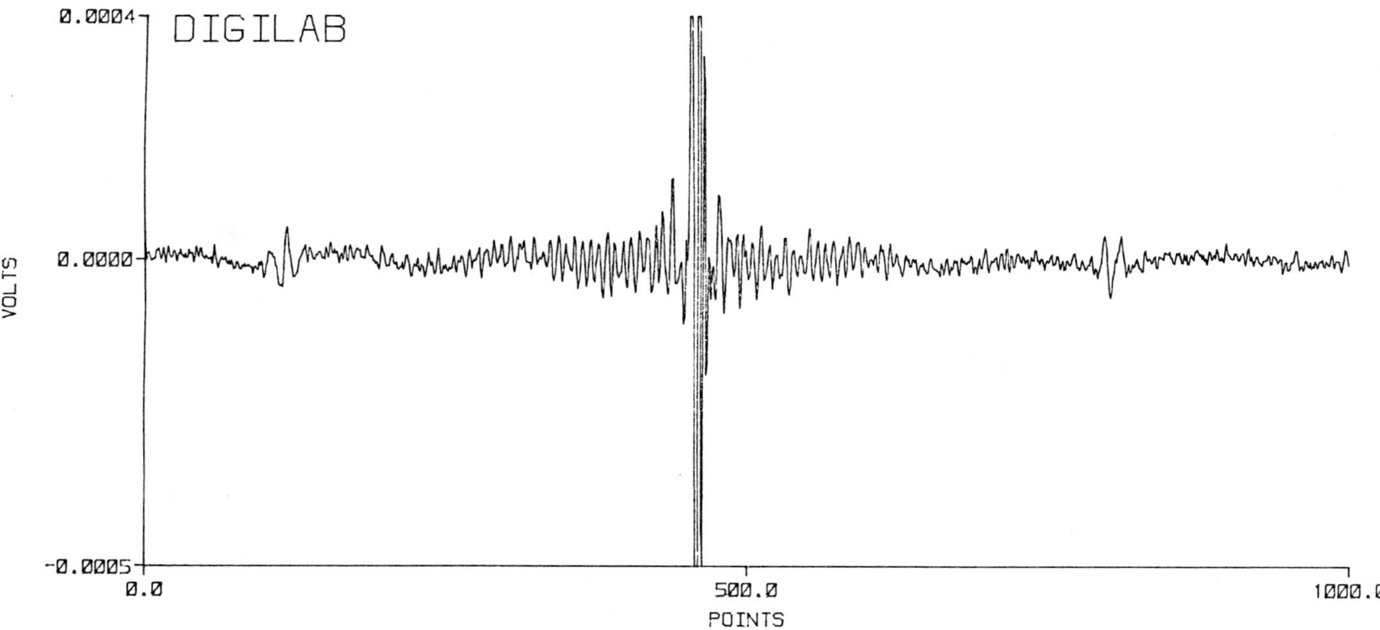

Figure 3. Interferogram from the 30 micrometer thick silicon epitaxial layer, with the sampling area set to 10x10 micrometers. 256 scans were coadded.

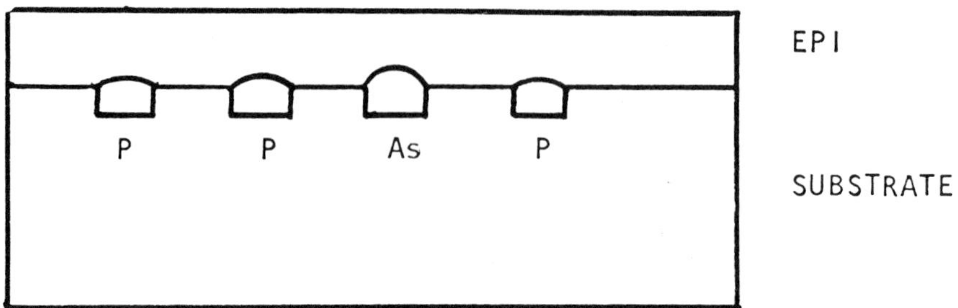

Figure 4. Schematic representation of a buried-layer epitaxial structure.

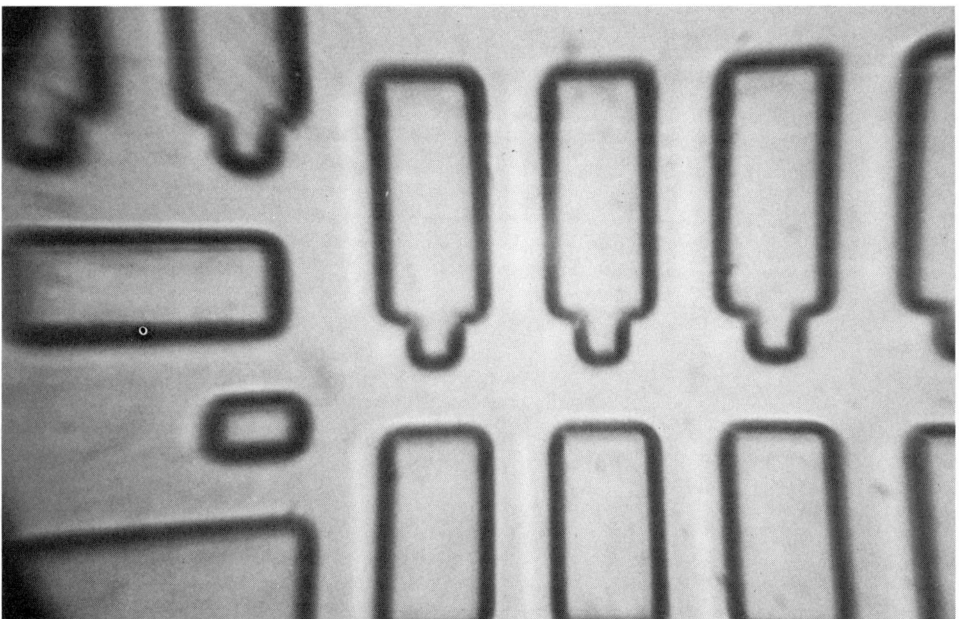

Figure 5. Photograph (360x magnification) of a silicon wafer showing the buried-layer patterns. The areas within the rectangular patterns are the buried-layers.

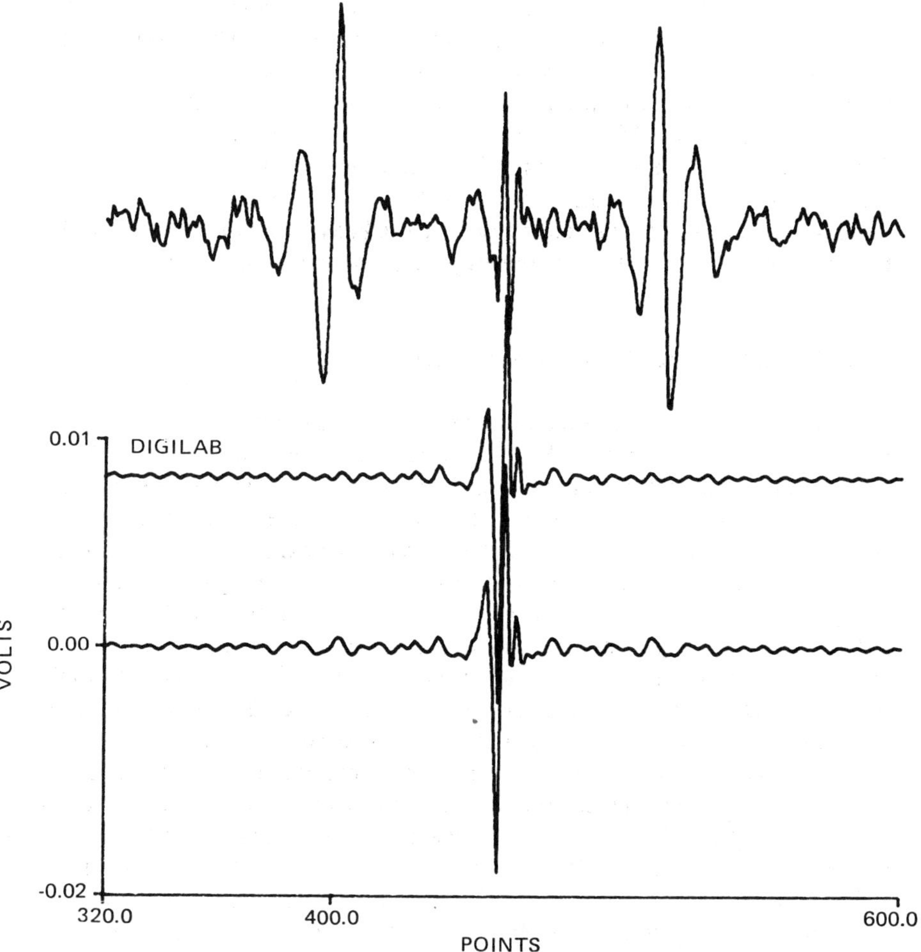

Figure 6. Interferograms from the silicon epitaxial buried-layer. Bottom - interferogram from the buried-layer; Middle - reference interferogram from an area adjacent to the buried-layer; Top - the difference interferogram clearly showing the sidebursts due to the buried-layer thickness.

One of the other applications of the FT-IR microsampling technique to the epitaxial measurements is the ability to map the epitaxial thickness over a whole silicon wafer using the computer-controlled motorized stage on the accessory. Figure 7 shows such a map from a 4 inch silicon epitaxy wafer (this is the same sample used for the measurements shown in Figures 2 and 3). Sixteen single scan measurements at 8 cm^{-1} resolution were made five millimeters apart over a diameter of the wafer with the sampling area set to 100 x 100 micrometers. One can clearly see the thickness variation over the wafer, the center of the wafer having a larger epitaxial thickness than the edges.

11. 3. 2. Oxygen and Carbon Measurements

The measurement of the interstitial oxygen and substitutional carbon concentrations in silicon using FT-IR spectroscopy is well documented [5]. These measurements are made by recording the infrared absorbance spectrum of the silicon sample of interest, and also that of an oxygen-, and carbon-free float zone silicon reference sample. A difference spectrum between the sample and the float zone spectra is then produced by cancelling out all the silicon phonon bands; this difference spectrum will show a band around 1107 cm^{-1} due to the interstitial oxygen, and a band around 605 cm^{-1} due to the substitutional carbon. The oxygen and the carbon concentrations are then computed from the measured absorption coefficients of these two bands. As the silicon devices are getting more and more densely packed, it is very important to be able to characterize these silicon properties on a microscopic scale. Krishnan and Kuehl [2] have shown the feasibility of such measurements. Kim and Smetana [6] have performed the oxygen concentration measurements from 100 x 100 micrometer sampling areas along an axial section on an AMCZ (axial magnetic field Czochralski grown) silicon. With the computer-controlled, motorized stage such inhomogeneity measurements in silicon can now be routinely performed. Figure 8 shows the infrared transmission spectrum from a 2 mm thick Czochralski silicon sample, with the aperture set to 100 x 100 micrometers.

Figure 9 shows an example of the measurement of the inhomogeneity of the oxygen concentration in a silicon sample. The sample used was a 2 mm thick, 10 cm long silicon crystal cut parallel to the growth axis. Measurements were made over 100 x 100 micrometer sampling areas five millimeters apart along the length of the sample. A float zone silicon sample spectrum, recorded over the same sampling size was used to eliminate the silicon phonon bands from all the recorded spectra. A plot of the resulting difference spectra in the oxygen region is shown in Figure 9. One can clearly see the discontinuity in

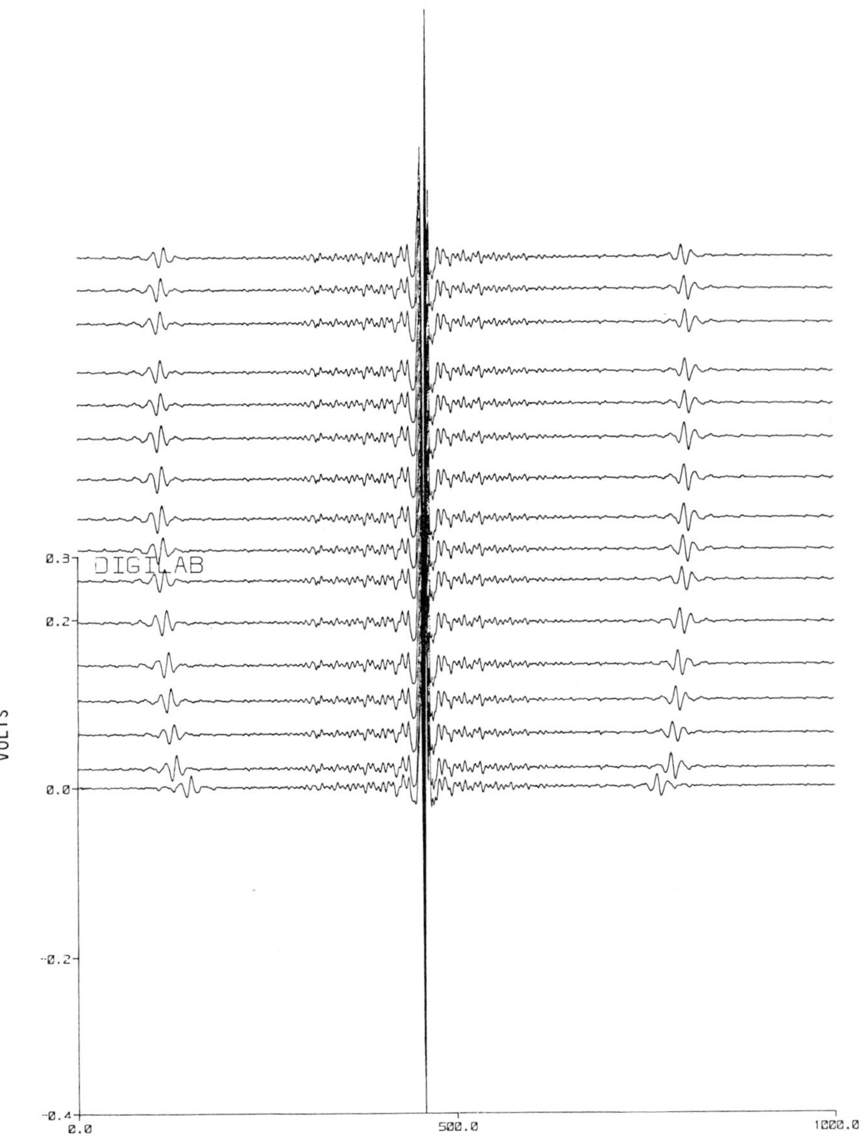

Figure 7. Variation of the epitaxial thickness on a 4" silicon wafer. The nominal thickness of the epi layer is 30 micrometers. The single-scan interferograms shown were recorded at 5mm intervals over one diameter of the wafer with the sampling area set to 100x100 micrometers.

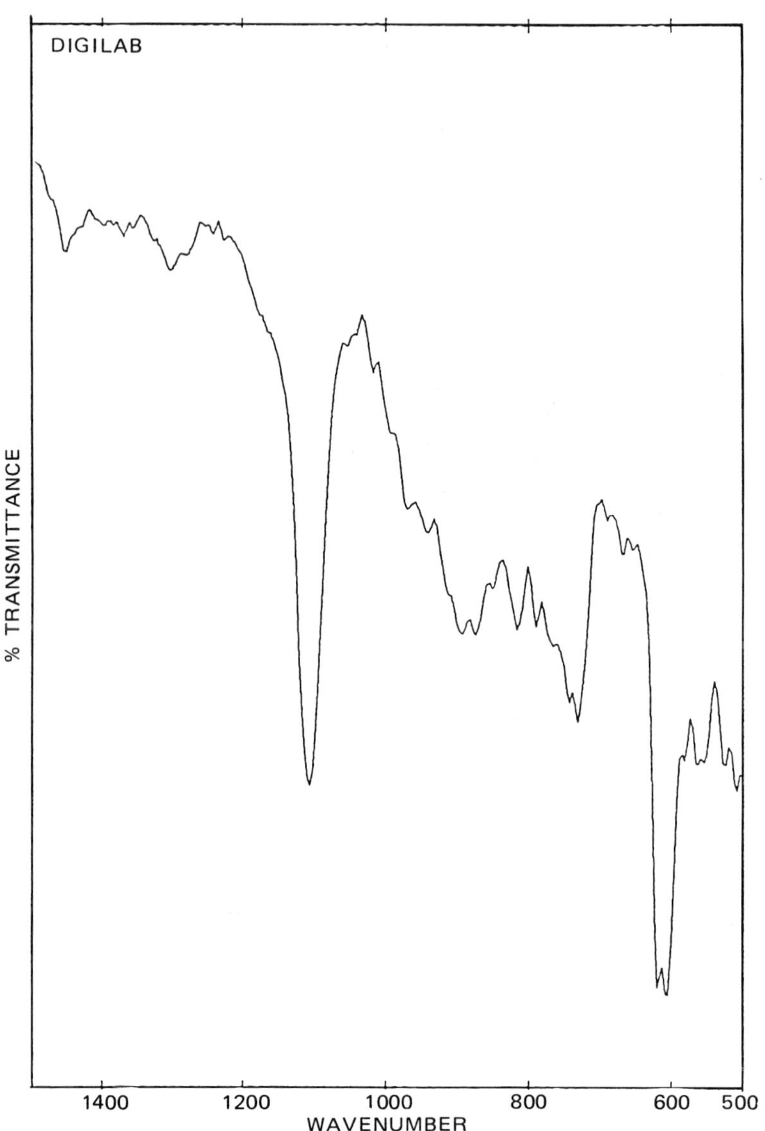

Figure 8. Infrared transmission spectrum of a 2mm thick silicon wafer containing interstitial oxygen. The sampling area was set to 100x100 micrometers, and the spectrum was recorded by coadding 128 scans at 8 cm^{-1} resolution.

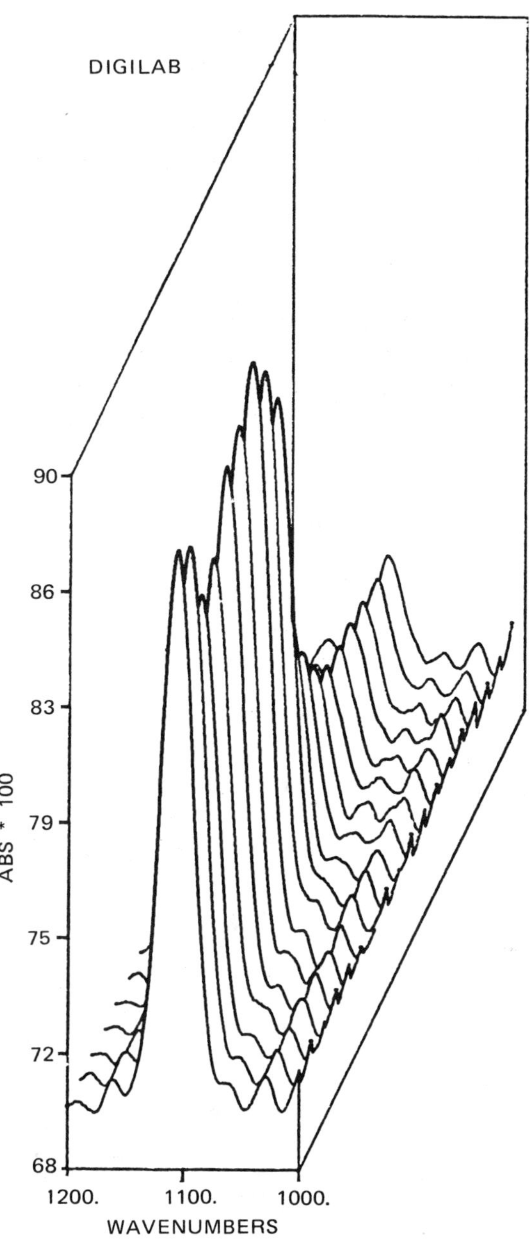

Figure 9. Difference spectrum showing the variation of the oxygen concentration along the growth axis of an AMCZ silicon sample. The measurement conditions were the same as for the spectrum shown in Fig. 8.

the oxygen concentration in the middle of the crystal. This result is similar to those presented by Kim and Smetana [6]. Measurements such as those shown in Figure 9 can be made on actual fabricated memory devices to measure the oxygen concentrations on a microscopic scale at different points on the device in order to better correlate the oxygen concentrations to the device performance.

11. 3. 3. Boron and Phosphorous Oxide Measurements

The FT-IR technique for the quantitative determinations of the concentrations of boron oxide and the phosphorous oxide in PSG (phosphosilicate), BSG (borosilicate), and BPSG (borophosphosilicate) glasses has been described in the literature [7]. In these systems, the major absorptions due to the B_2O_3, and

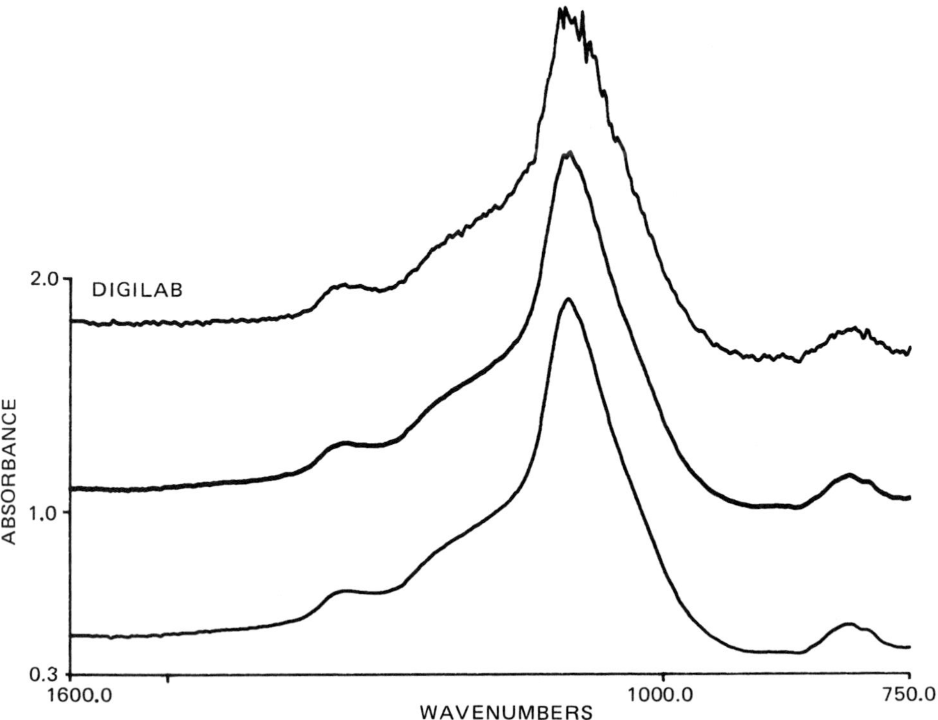

Figure 10. Infrared absorbance spectra of a PSG layer on silicon. From bottom to top the sampling area was set to 100x100, 50x50, and 25x25 micrometers, respectively. The absorption band due to the P_2O_5 in the PSG layer can be seen at around 1330 cm^{-1}.

the P_2O_5 moeities appear around 1400, and 1330 cm^{-1}, respectively. With the microsampling accessory in the transmission mode, these measurements could now be made from extremely small areas of the sample. Figure 10 shows the FT-IR transmission spectra of a PSG film containing around 4 weight % phosphorous recorded with the sampling areas set to 100 x 100, 50 x 50, and 25 x 25 micrometers. In all these spectra one can clearly see the band around 1330 cm^{-1} due to the phosphorous - oxygen stretching vibration. The fact that such spectra could be obtained from such small areas opens up the possibility of measuring these properties of the glass films on finished silicon device wafers containing buried layers, metallization etc.

11. 4 CONCLUSION

It has been shown that a number of silicon properties during various stages of device manufacturing can be characterized on a microscopic scale from a variety of transmission and reflection measurements. These measurements will be of great importance as the silicon devices are getting more and more densely packed.

REFERENCES

1. K.Krishnan and R. Mundhe, *Spectroscopic Characterization Techniques for Semiconductor Technology,* in *SPIE Proceedings*, Billingham, WA, Vol. 452, (1983).
2. K. Krishnan and D. Kuehl, in *Semiconductor Processing* (Dinesh C. Gupta, ed.), ASTM STP 850: 325 (1984).
3. K.Krishnan, *Polymer Preprints,* 25: 182 (1984).
4. K. Krishnan, *Optical Techniques for Industrial Inspection*, in *SPIE Proceedings* (P. G. Cielo, ed.), Billingham, WA Vol. 665: 252 (1986).
5. K. Krishnan, in *Proceedings of the Symposium on Defects in Silicon* (W.M. Bullis, and L. C. Kimerling, eds.), The Electrochemical Society Inc, Pennington, N. J. Vol. 83-9: 285, (1983).
6. K. M. Kim, and P. Smetana, *J. Electrochem. Soc.*, 133: 1682 (1986).
7. K. Krishnan, in *Semiconductor Processing* (Dinesh C. Gupta, ed.), ASTM STP 850: 325 (1984).

Application of Infrared Microspectroscopy to Biological and Pharmaceutical Research

12

Biological Applications of FT-IR Microscopy

MICHAEL P. FULLER AND ROBERT J. ROSENTHAL *Spectroscopy Research Center, Nicolet Instrument Corporation, Madison, Wisconsin*

12.1 INTRODUCTION

The combination of Fourier transform infrared (FT-IR) spectrophotometers and various microsampling devices is dramatically changing life for the spectroscopist. Perhaps the most important device for microsampling is the latest generation of FT-IR microscopes [1]. This accessory permits routine investigations of materials thought unsuitable for infrared study a short time ago [2-5].

An area of increasing emphasis is the application of infrared spectroscopy to the life sciences. The combination of FT-IR with microscopy is capable of elucidating questions related to the basic tenets of the science. This powerful instrumentation allows the spectroscopist to "see", now in a chemical sense, inside various biological materials. Our results from a varied selection of materials represent only a small sample of what can be accomplished. It is our hope that this will foster further investigations by other spectroscopists.

12.2 EXPERIMENTAL

The majority of the work described was performed using the Spectra-Tech (Stamford, CT) Spectra-Scope. The Spectra-Scope is a small, easy to use, relatively inexpensive infrared microscope. It is seen in Figure 1 mounted in a Nicolet 5DXC auxiliary module with the camera and monitor option. The

Figure 1. Spectra-Tech Spectra-Scope with video camera and monitor mounted in 5DXC spectrophotometer auxiliary module.

monitor option is very useful with this particular microscope because it provides a heads-up type display allowing easy control of the X,Y,Z stage controls while viewing the magnified image of the sample. The monitor on this particular slide shows the image of the end of a human hair. Several spectra were obtained using a different system, the Spectra-Tech IR-plan microscope. The IR-plan Microscope is seen in Figure 2 coupled with a Nicolet 5DXC spectrophotometer using a specially designed interface module. Very high quality spectra are obtained with either of these two microscopes. In all cases a high sensitivity MCT detector was employed with the 5DXC instrument operating at 4 or 8 wavenumber resolution. Typically 32, 64 or 128 scans were co-added to yield the resulting spectra.

Figure 2. Spectra-Tech IR-plan microscope coupled with Nicolet 5DXC spectrophotometer.

12.3 RESULTS AND DISCUSSIONS

A large number of the spectra described involve measurements performed on plant structures. One of the things that is apparent from the spectra of Figure 3 is that a leaf is not necessarily just a leaf. There are distinctly different physical regions in a leaf and Figure 3 shows the absorbance spectra of a vein region and the symplast or tissue region of a lettuce leaf [6]. These spectra are obviously quite different. Subtraction of the spectrum of the symplast region from that of the vein results in the spectrum shown in Figure 4. The broad band around 1100 wavenumbers due to cellulosic structures is found to be more highly concentrated in the vein portion of the leaf relative to that of the symplast

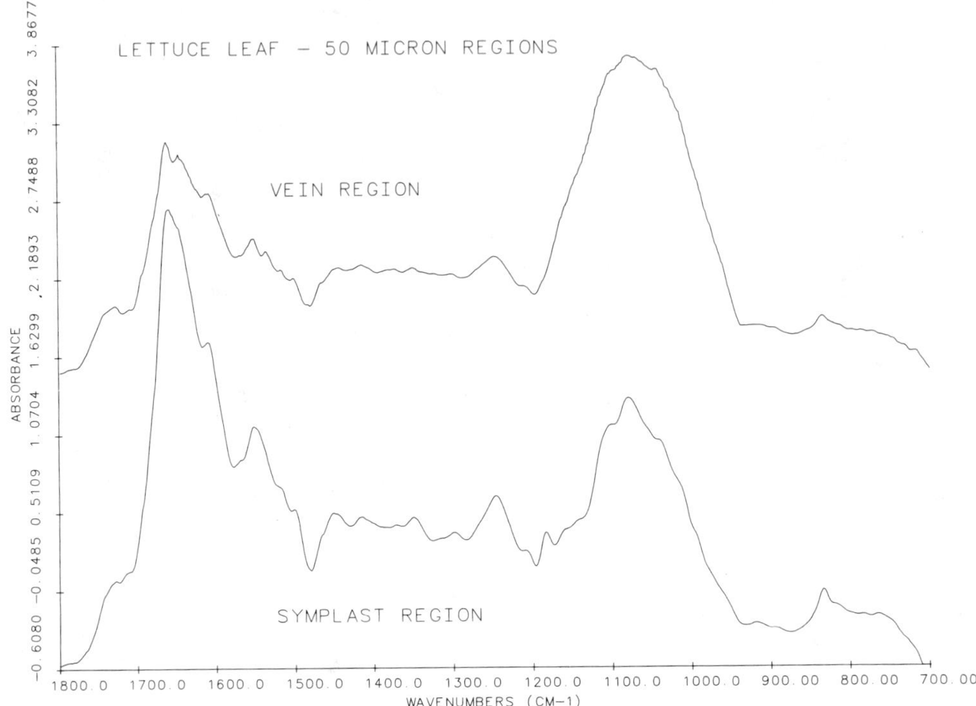

Figure 3. Spectra of the vein (top) and symplast (bottom) regions of a lettuce leaf. The area measured in each case was about 50 x 50 microns.

region. It is also apparent that the relative water concentration is higher in the symplast region as evidenced by the 1640 cm^{-1} water absorption being in the negative direction.

Because the infrared beam is highly focused by the microscope, care must be taken when measuring the spectra of volatile samples. The first leaf spectrum obtained, when running a series of different leaf samples, showed peak absorbances that were in the range of two to three absorbance units. This is not unexpected since leaves have a high water concentration and the ones being measured were quite thick. Even so, it was still a very useful spectrum. When the microscope stage was moved to another portion of the leaf the absorbances throughout the spectrum were considerably higher than four absorbance units making the spectrum unusable. This was perplexing and disturbing considering the excellent quality of the first spectrum and the fact

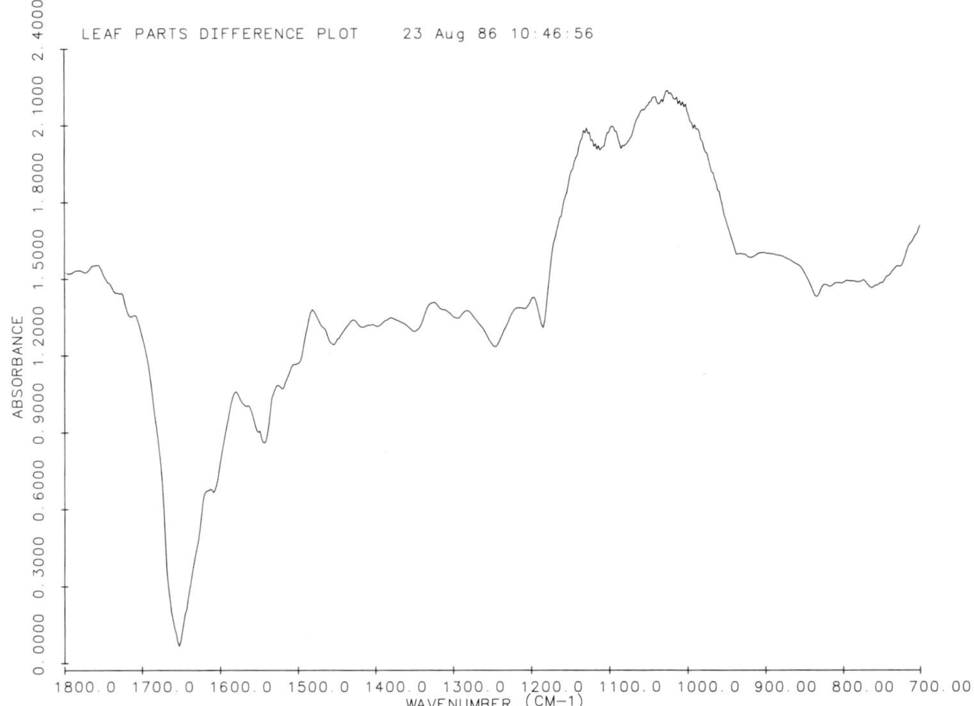

Figure 4. Difference spectrum obtained by subtracting the symplast region spectrum from that of the vein.

that the aperture was left unchanged. Similarly poor results were obtained when analyzing a number of different regions on different leaves.

In an attempt to improve the signal-to-noise ratio of the data at these high absorbance levels 1000 scans were co-added. During the data collection, about half way into the scan, the analog-to-digital converter began overflowing. This was unexpected since the gain setting was proper when the data collection was initiated. The scan was restarted at a lower gain and the spectrum obtained compared well with the first (Figure 3) spectrum recorded. Comparing these results with the intermediate ones it was determined that water was being evaporated from the leaf over time by the highly focused infrared beam. This resulted in lower absorbance values and better spectra. There is a considerable amount of energy in a very small area when using the infrared microscope and over the course of about ten minutes, about 60% of the water in the small region

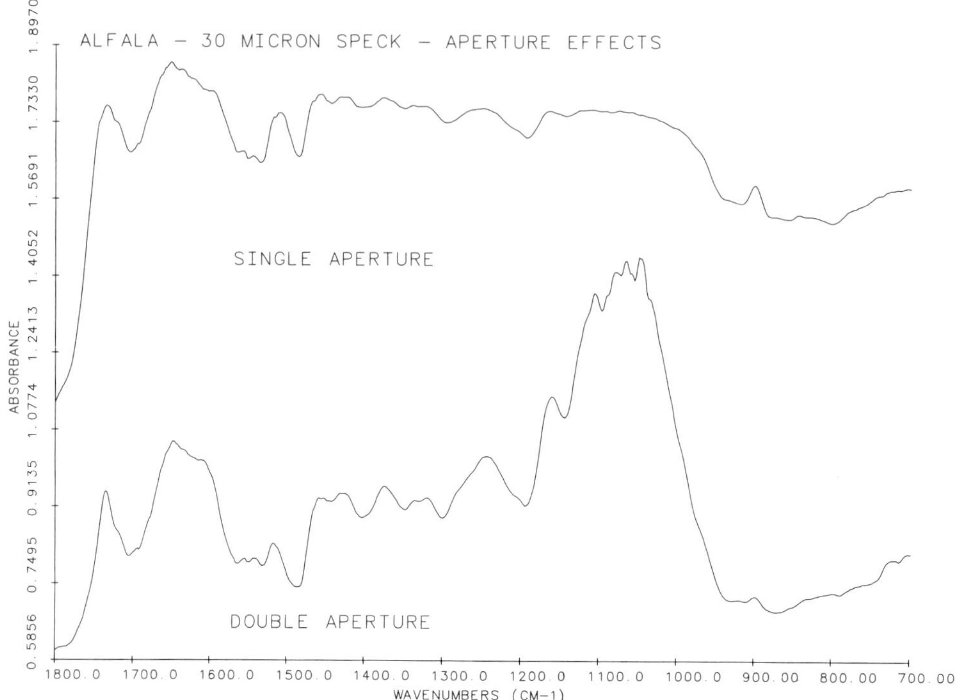

Figure 5. Top - Result obtained using a single aperture to measure the spectrum of a speck of alfalfa leaf. Bottom - Improved results obtained using the redundant apertures to the IR-plan microscope.

upon which the infrared radiation impinges is evaporated. Since this is dehydration on a micro scale, the water may return to the site from the surrounding tissue when the beam is removed. This may provide a probe for studying dehydration in plant as well as animal cells. In the remainder of the spectra described, reasonable care was taken so that the samples were in contact with the focused beam for approximately the same length of time in each case.

The effect of stray light on spectra of highly absorbing samples is well known [7]. When using the infrared microscope it is important to eliminate, as much as possible, light passing around the sample rather than through it [8]. The spectra of Figure 5 result from the measurement of a piece of ground alfalfa, commonly used as a supplementary feed for cattle. A tiny speck of this ground alfalfa was placed on the stage and the top aperture closed around it, using the IR-plan microscope, to produce the top spectrum of Figure 5. It is apparent especially at low frequencies, or longer wavelengths, that there are severe distortions in the bands due to stray light. If that same speck of alfalfa is analyzed using redundant apertures (Figure 5 - bottom) the stray light effect is greatly reduced, yielding more closely the true band shape of the very strong cellulose band around 1100 wavenumbers. Therefore, selecting the proper apertures is very important when running samples where one is concerned about the effects of stray light.

There are definite characteristics of a particular alfalfa which make it better or worse than others as a food supplement for animals [9]. One factor related to feed quality is plant age. There is an optimal cutting time during the plant growth cycle which yields the highest quality feed. Figure 6 shows the spectra obtained from the tips of leaves from an old plant and a young plant. Subtracting the spectrum of the older plant from that of the younger, the difference plot of Figure 7 is obtained. This spectrum suggests that there is a difference in the cellulosic nature of the two samples, with the older plant having a higher relative cellulose content. There are also carbonyl absorptions around 1700 wavenumbers indicating a higher concentration in the older plant. A higher protein content concentration in the young plant versus the older one is suggested by the downward going Amide I and Amide II bands at 1650 and 1545 wavenumbers.

During a recent herbicide study, the question of interest was whether or not a number of small brown spots on a series of birch leaves were due to misdirected herbicide or were the result of insects taking bites out of the leaf causing dead spots. A sample of the leaf was measured, yielding the spectrum of one of the dark spots and that of a control area as shown in Figure 8. Because of the thickness of the leaf the resulting absorbances are, in many spectral regions, well above two units. Although the sample is quite strongly absorbing, it was still possible to obtain useful infrared spectra. Subtracting the spectrum of a control leaf and a spotted leaf, a number of features become apparent in the difference plot of Figure 9. Although there is some change in the cellulosic areas around 1100 to 1250 wavenumbers, the bands around 1700 wavenumbers indicate oxidation of the leaf rather than herbicide contamination. This suggests that this particular spot was a result of simple oxidation of the leaf from a cutting of the plant tissue, probably related to insect feeding.

On the other hand, when examining the spectra obtained from samples known to have been treated, at a prior time, with an herbicide and then exposed to wet and dry cycles in a weatherometer (Figure 10), distinct evidence suggesting the presence of the herbicide is seen. Subtracting the spectrum of a control birch leaf from that of the treated (Figure 11) shows, very clearly in this

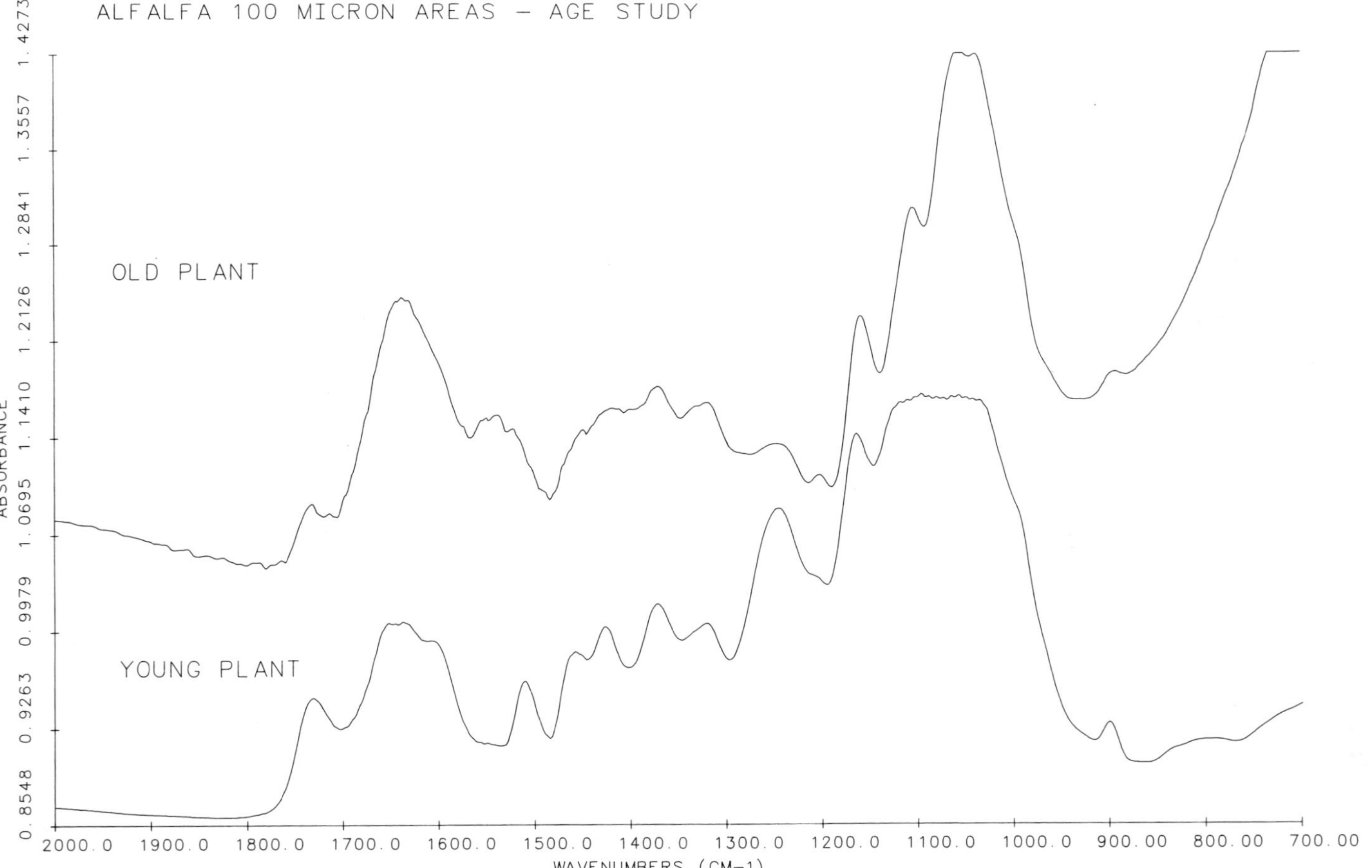

Figure 6. Top - Spectrum obtained from 100 x 100 micron section of an old alfalfa leaf. Bottom - Spectrum from same size area of young alfalfa leaf.

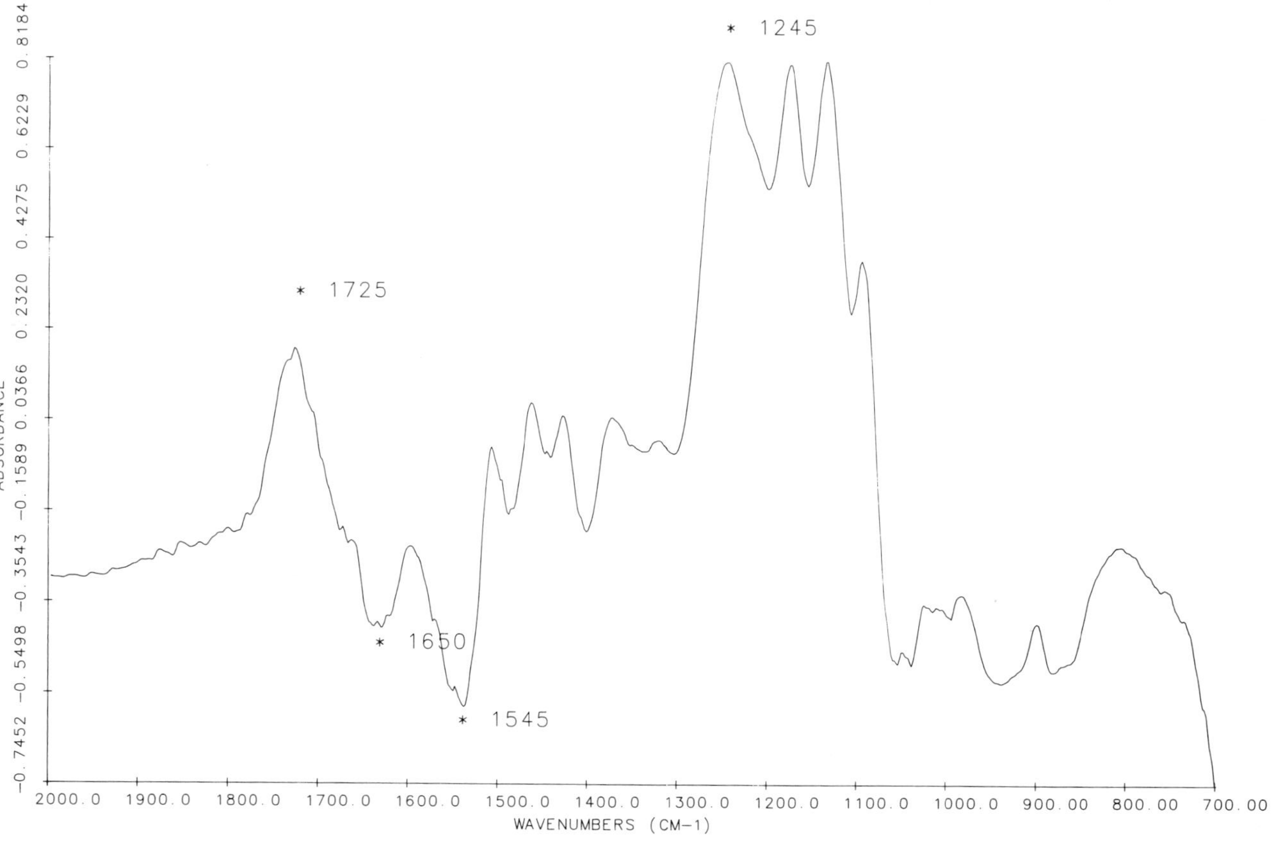

Figure 7. Difference spectrum obtained by subtracting the spectrum of the young plant from that of the older one. Upward going bands indicate a higher relative concentration in the older leaf and downward a higher concentration in the younger plant.

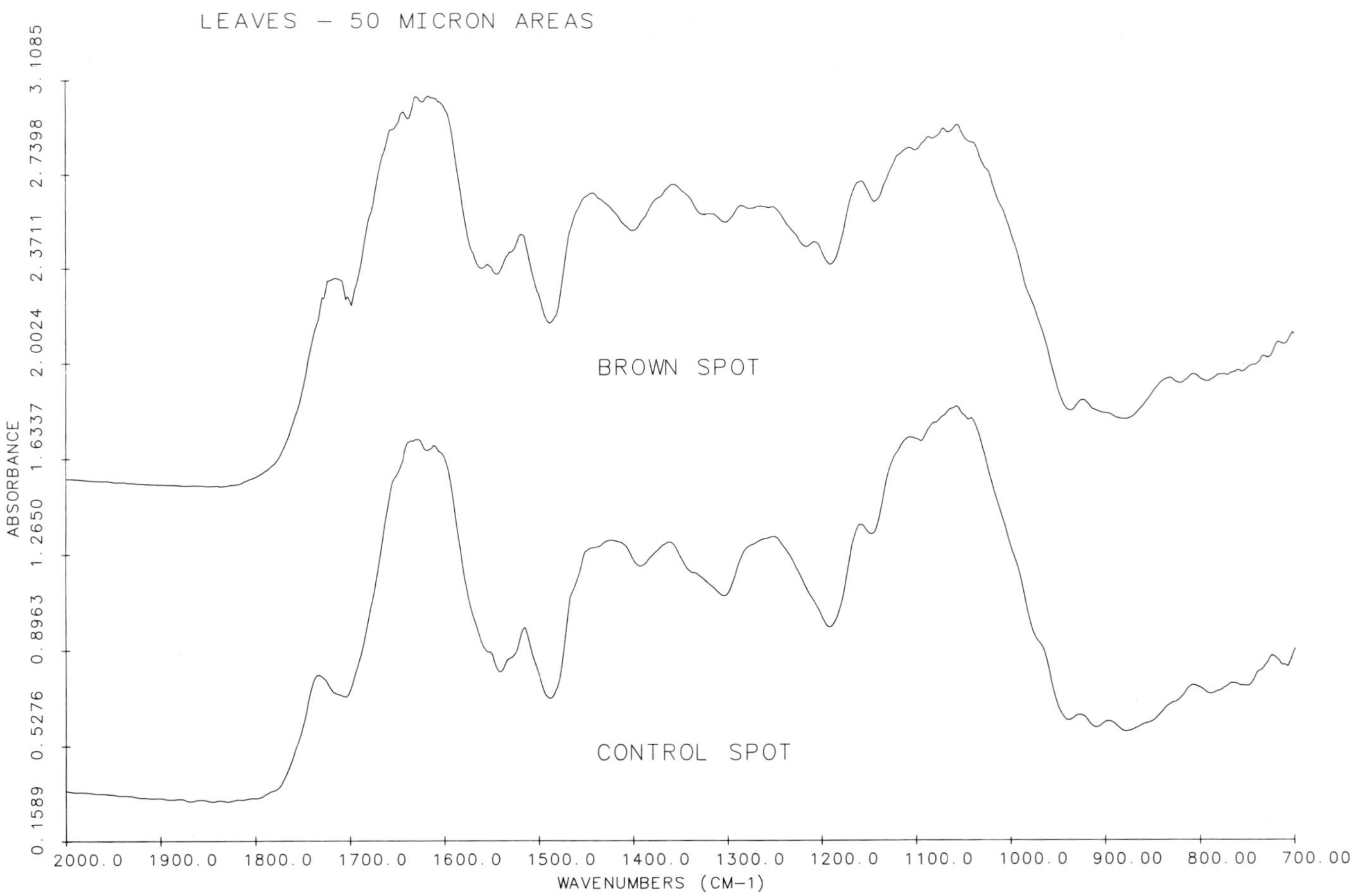

Figure 8. Spectra from 50 x 50 micron areas of: top - brown spot on birch leaf and bottom - a control leaf.

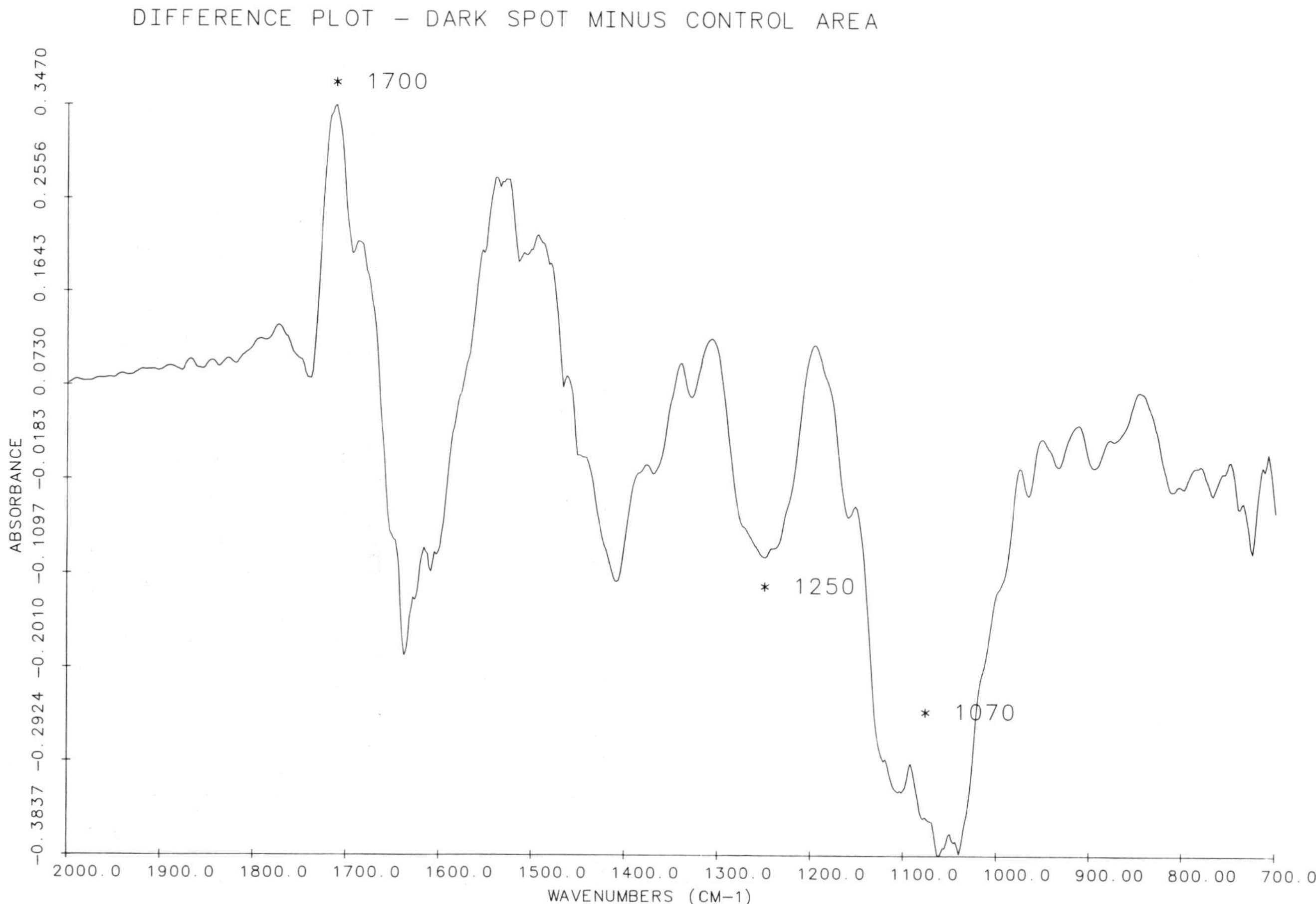

Figure 9. Difference spectrum obtained by subtracting the spectrum of the control leaf from that of the brown spot.

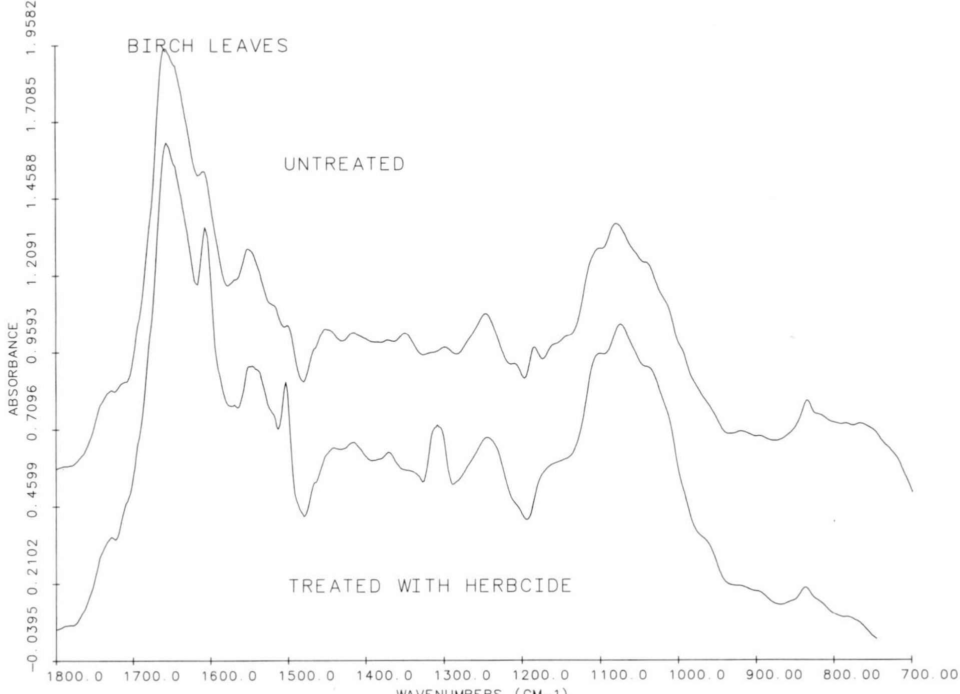

Figure 10. Top - Spectrum of an area of a control leaf. Bottom - Spectrum obtained from a discolored spot on a birch leaf.

case, bands assignable to characteristic aromatic features around 1500 and 1600 wavenumbers. These features, along with the absorption at approximately 1300 wavenumbers, correlate well with the spectrum of the particular type of herbicide used in this study.

Guar beans are ground, extracted and used as a thickening agents in, among other things, ice cream and toothpaste. The beans are often treated with an alkyl oxide gas prior to being ground and extracted, to give the thickening agent. The goal of this experiment was to analyze a microtomed section of treated bean to determine the depth of penetration and reaction of the gaseous oxide as a means of optimizing the treatment conditions. A thin microtomed section, cut through the thickest part of the bean, was placed on a barium

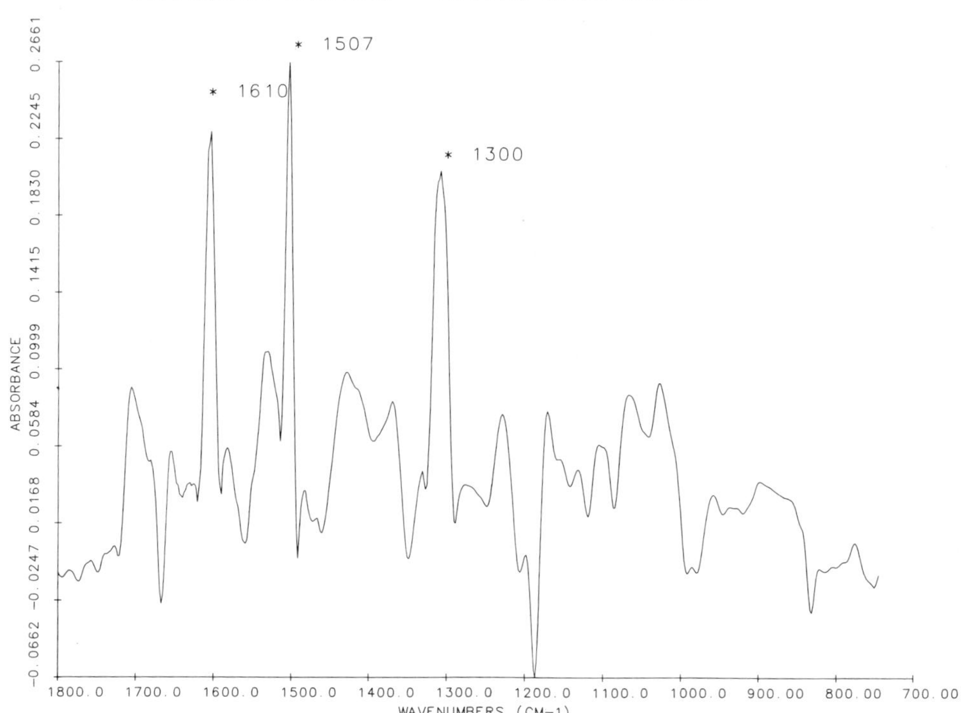

Figure 11. Difference spectrum obtained by subtracting the spectrum of the control leaf from that of the discolored area. The bands marked are assignable to the herbicide used to treat the leaf.

fluoride plate and then onto the stage of the infrared microscope. Spectra at the edge of the bean and toward the center of the slice of the bean were measured using the microscope and are presented in Figure 12. After subtracting these two spectra the resulting difference spectrum (Figure 13) shows a number of absorptions that can indeed be related to the degree of reaction of the cellulosic backbone of the bean with the oxide gas to produce new ether (-C-O-C-) linkages, as evidenced by the band around 1175 wavenumbers. The absorptions around 2950 wavenumbers are assignable to the C-H stretching modes of aliphatic groups added as a result of the reaction. The extent of reaction as a

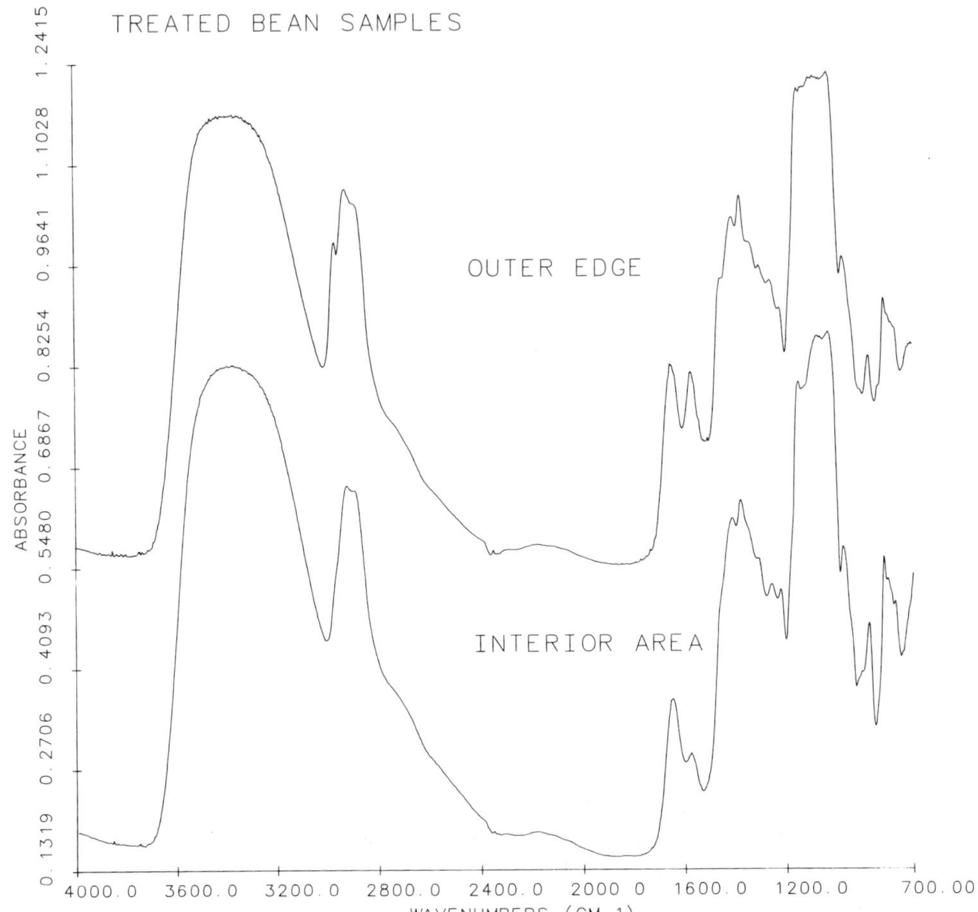

Figure 12. Top - Spectrum from the outer edge of a microtomed section of a guar bean. Bottom - Spectrum from the interior region of the same section.

function of depth into the bean is determined by simply moving the microtomed section on the microscope stage and making multiple measurements.

Another potential use of infrared microscopy is to measure the extent to which different parts of cells uptake stains or various chemicals [10]. In this case, a plant cell treated with a typical plant cell stain and an unstained control

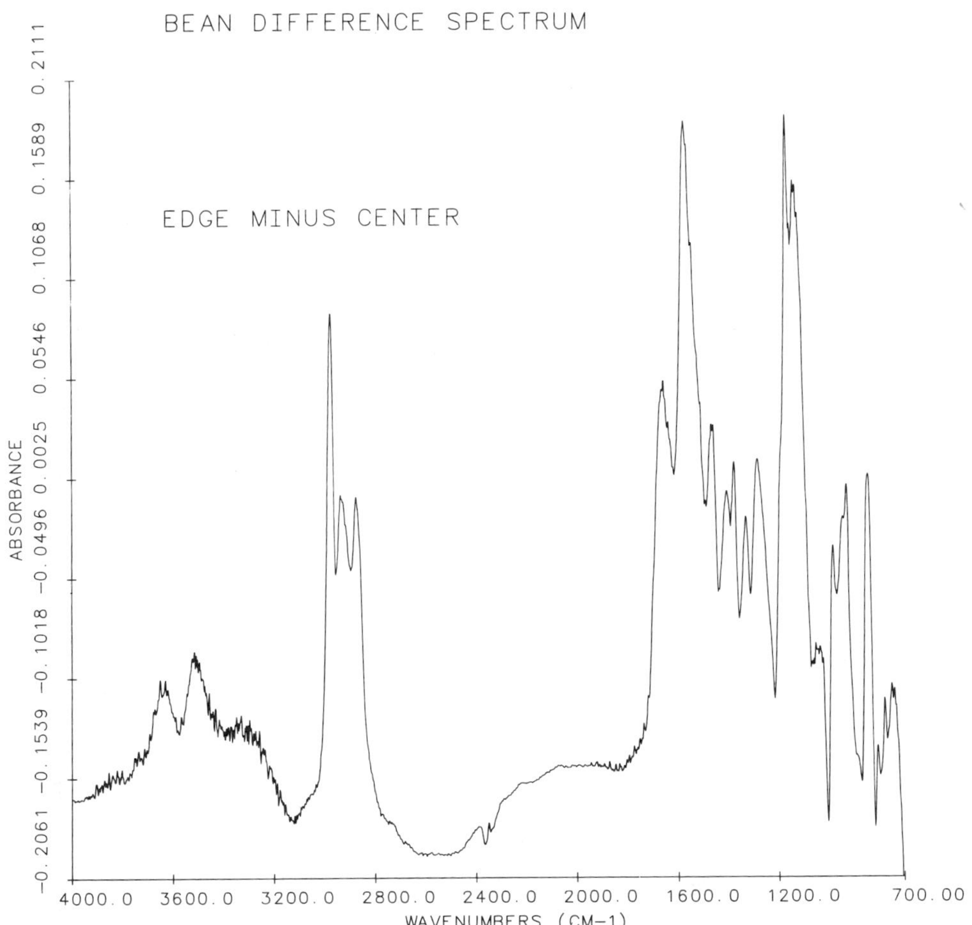

Figure 13. Difference spectrum obtained by subtracting the spectrum of the interior of the treated guar bean from that of the outer area. This spectrum shows quite clearly the chemical changes produced in the bean as a result of the treatment.

cell were analyzed (Figure 14). After a subtraction (Figure 15), one can identify a number of bands which are indeed related to the uptake of the stain into the plant cell. Differences between this plot and the spectrum of the neat stain can be related to chemical associations between the stain and the plant

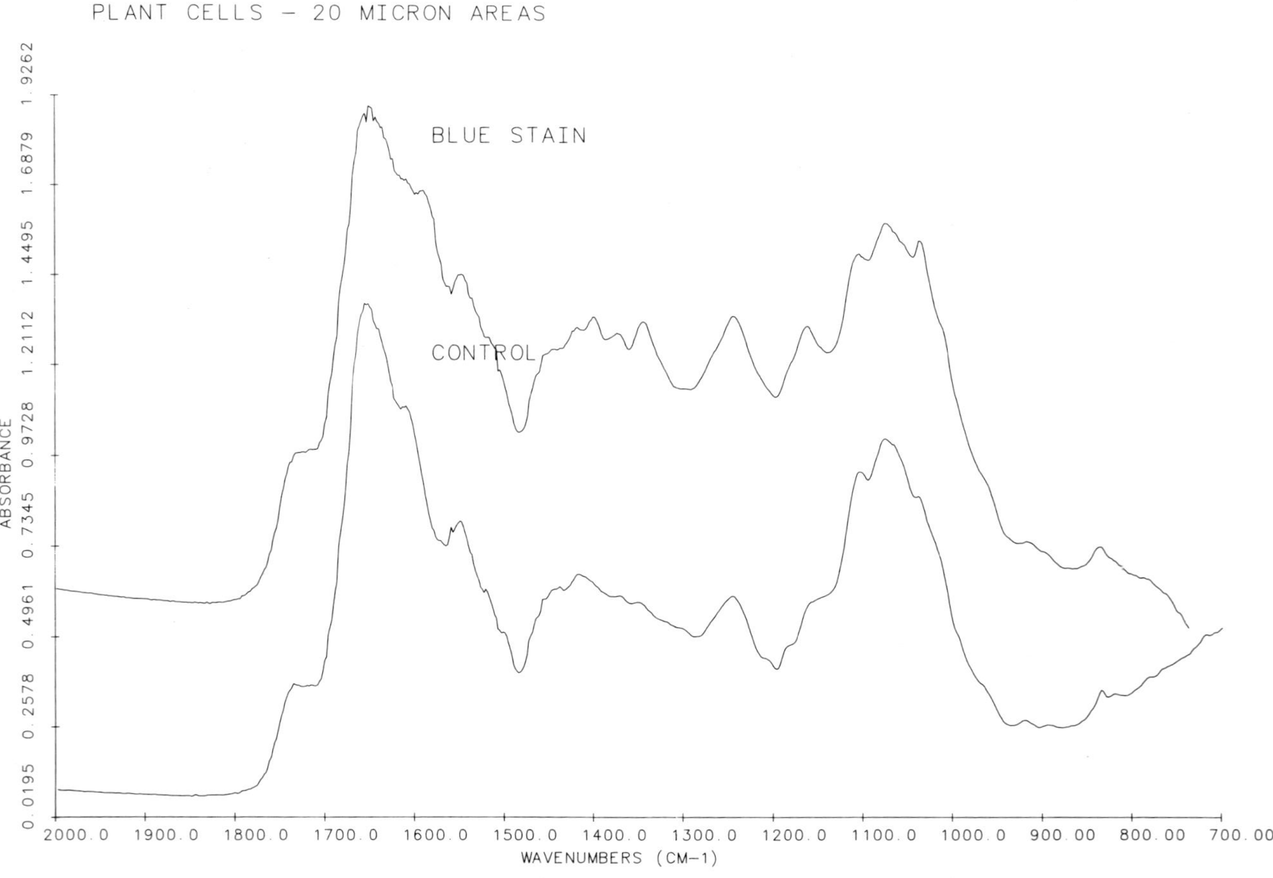

Figure 14. Top - Spectrum from an area of approximately 20 x 20 microns of a plant cell stained with an experimental halogenated alkylphenol dye. Bottom - spectrum from an area 50 x 50 microns square of a control plant cell.

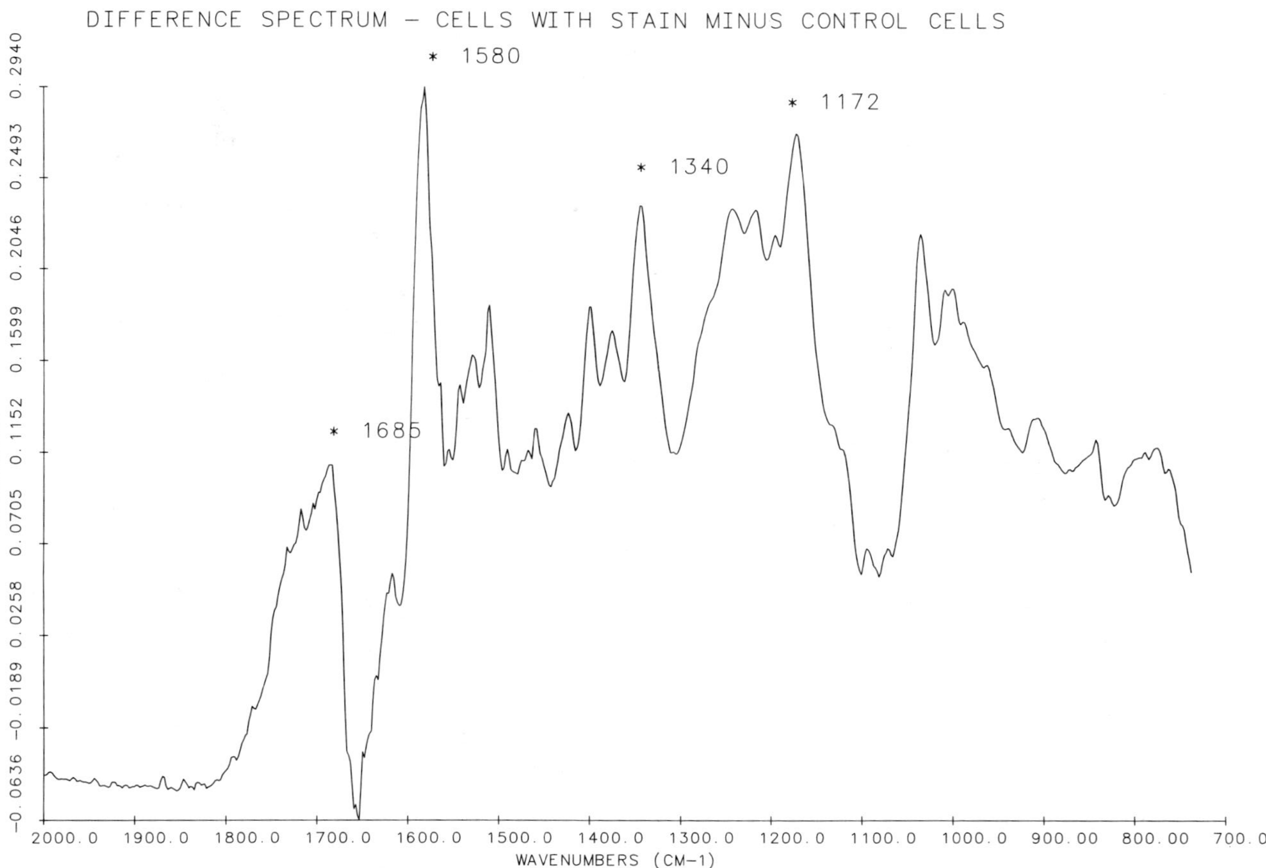

Figure 15. Difference spectrum obtained by subtracting the spectrum of the control cell from that with the stain. The indicated bands are assignable to aromatic (1580 cm^{-1}), methyl (1340 cm^{-1}) and phenol (around 1172 cm^{-1}) absorptions.

cells. Again, this is simply an indication of what can be easily accomplished using the IR microscope.

Skin samples are often used by health and beauty care product companies to study the interactions of various chemical mixtures with living tissue [11]. IR microscopy offers a valuable tool for these types of analyses. In this case a piece of living skin was treated with a conditioning emollient and then subjected to various washes. The purpose of this experiment was to determine what components of the lotion remained after the wash cycles. Living skin is relatively expensive and difficult to come by. Obviously, the smaller the area that is used, the more experiments that can be performed on a given piece of flesh. It is therefore useful to use very small samples. The infrared microscope offers a tool for analyzing these small samples. The specimens can be mounted on barium fluoride, germanium or zinc selinide crystals prior to analysis using the infrared microscope. A piece of treated skin and a control sample (both areas of about 200 microns) were measured yielding the spectra shown in Figure 16. The resulting difference spectrum (Figure 17) shows a band around 730 waven umbers due to the long chain aliphatic hydrocarbon portion of one the lotions components, while the 1040 band is due to a hydroxy C-O stretching mode related to one of the other active ingredients in the emollient. This example demonstrates how the infrared microscope can be used as a sensitive probe for measuring the presence and effects of chemicals on skin.

Infrared spectroscopy can be used to determine a number of the structural characteristics of protein samples [12]. Spectra obtained from human blood can be very interesting. The spectra of fresh red blood cells five minutes after deposition onto a barium fluoride plate and then after twenty minutes are presented in Figure 18. Superficially these spectra appear identical. If you subtract these two spectra (Figure 19 , however, significant differences in the -OH stretching region around 3200-3700 wavenumbers are observed. These differences don't correlate with the changes in the 1640 wavenumber water absorption band. This suggests a change in the hydrogen bonding and molecular conformations of the protein occurring as a function of water content which changes as a result of evaporation over time. The expanded plot of the difference spectrum, shown in Figure 20, indicates that quite a large change in the protein Amide-I and Amide-II absorptions has occurred. These changes are similar in appearance to some of the spectral differences in protein spectra described by other authors and may be related to conformational changes of the blood proteins [13].

An attempt was made to obtain the spectrum of a single human red blood cell using the IR-plan microscope. Single blood cells are 8.5 microns in

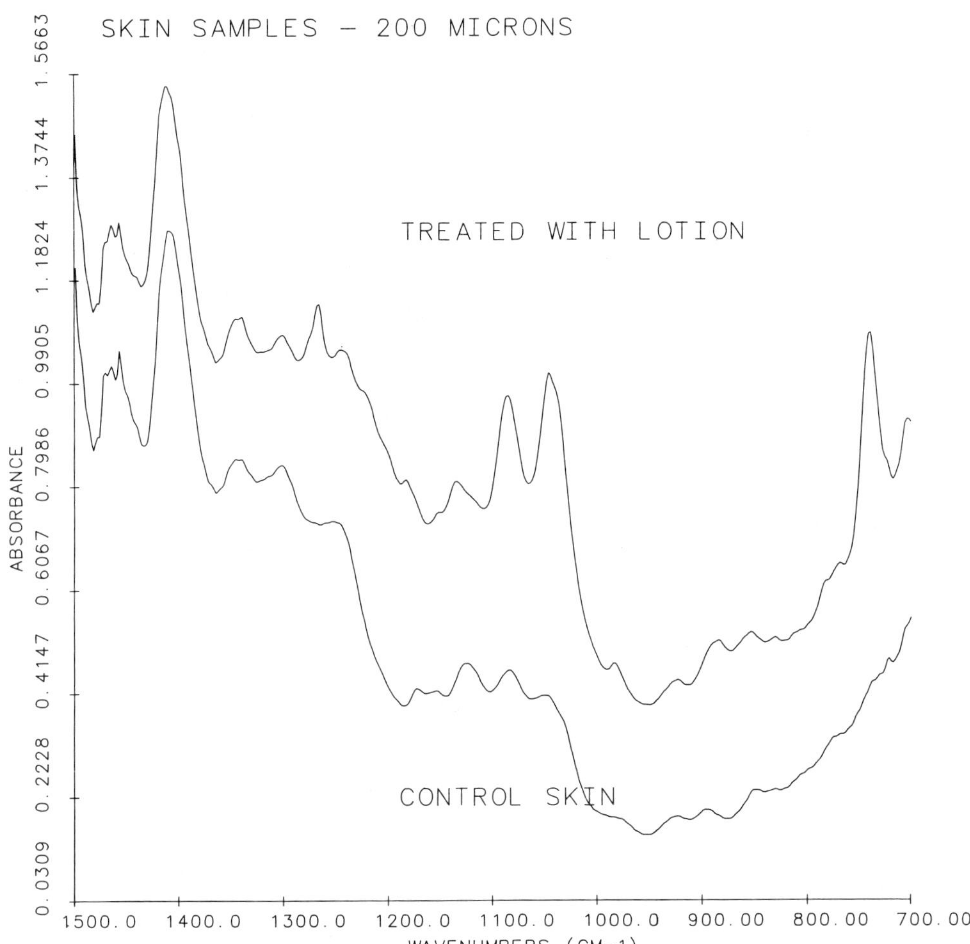

Figure 16. Top - Spectrum of living human skin (200 x 200 microns) after treatment with an emollient followed by a number of wash cycles. Bottom - Spectrum of a piece of control skin (200 x 200 microns).

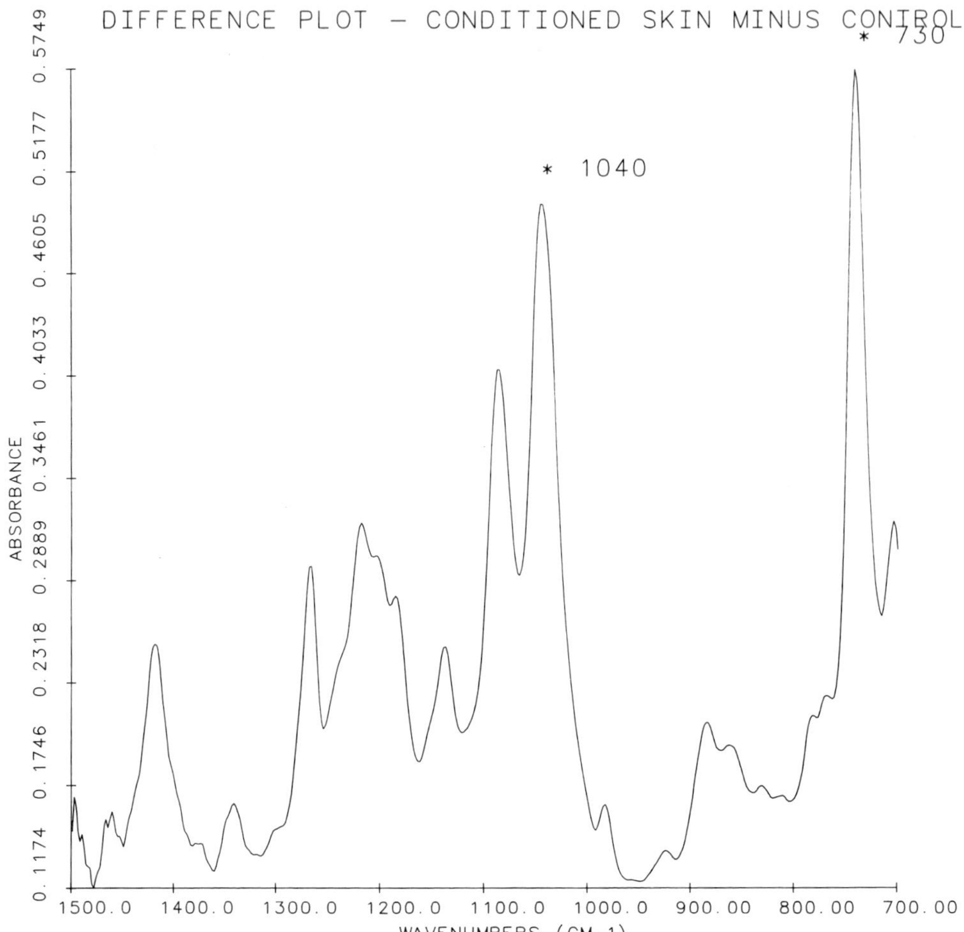

Figure 17. Difference spectrum obtained by subtracting the spectrum of the control skin from that which had been treated then washed. The bands around 730 and 1040 wavenumbers are assignable to components of the emollient.

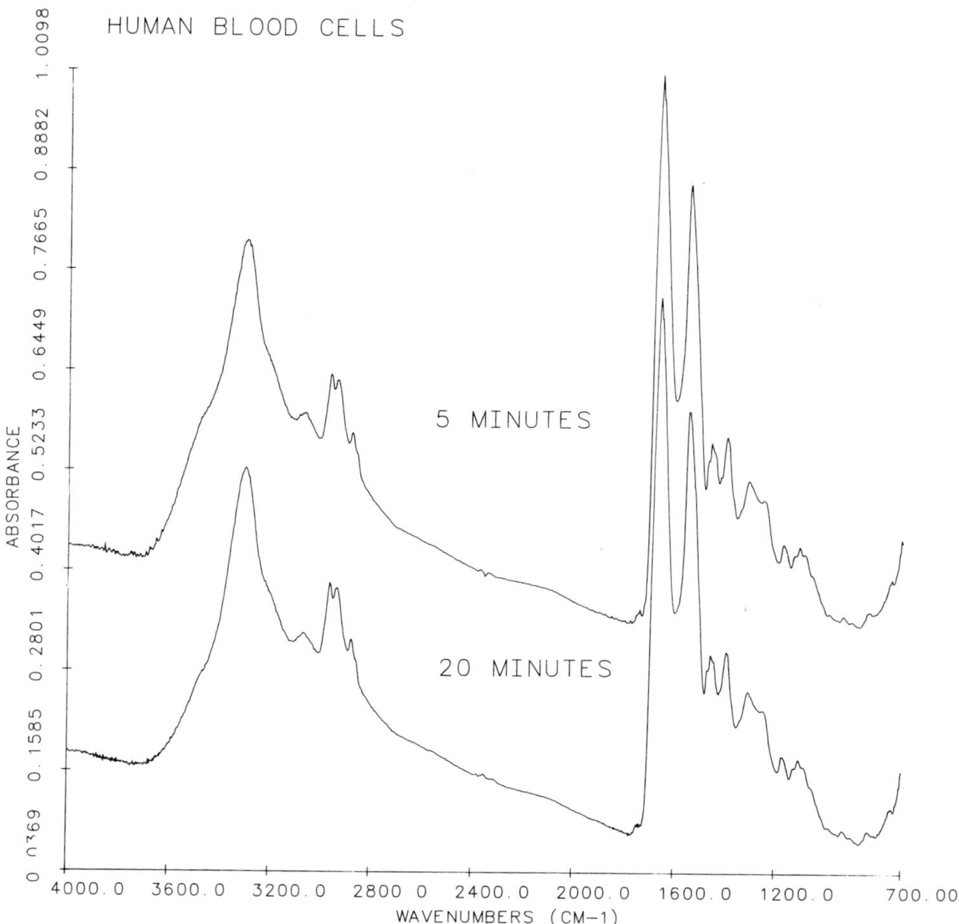

Figure 18. Top - Spectrum of human blood five minutes and Bottom - Spectrum after twenty minutes under the infrared beam of the microscope.

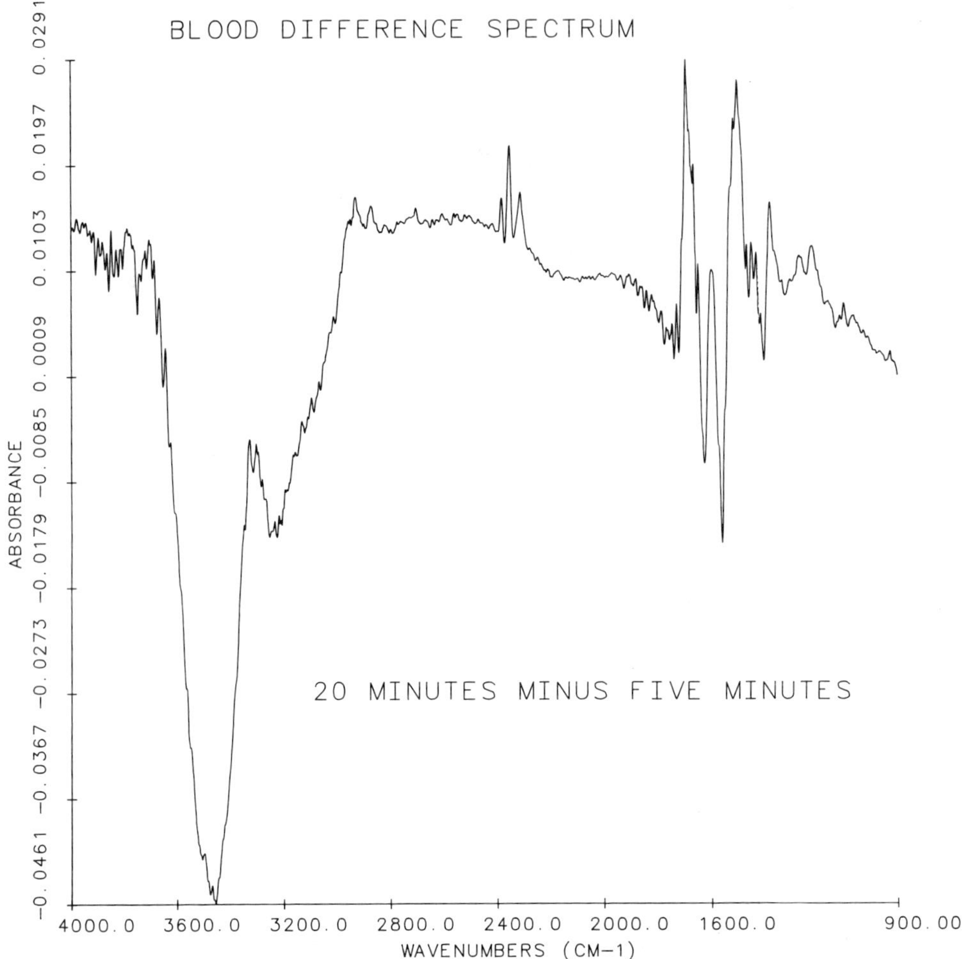

Figure 19. Difference spectrum obtained by subtracting the spectrum of blood after five minutes from that after twenty minutes.

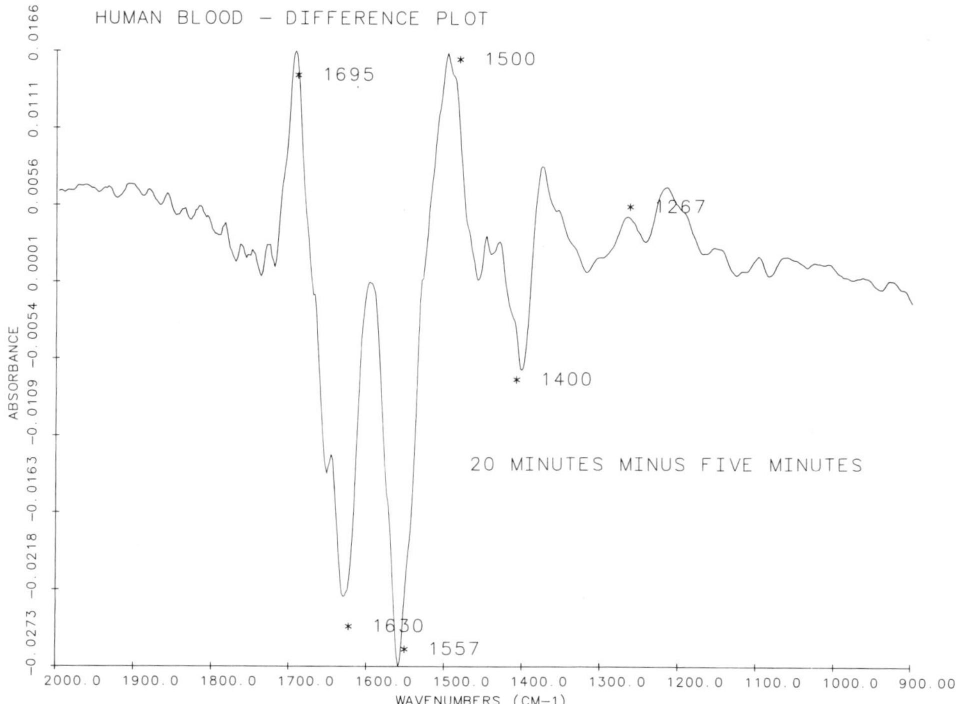

Figure 20. Expanded difference spectrum of Figure 19 showing the carbonyl region. The effect of conformational changes is seen in the Amide-1 and Amide-II absorbance residuals.

diameter by about 1.5 microns thick and are actually used as a size standard because of their uniformity. A usable spectrum of a single red blood cell could not be obtained, but a spectrum of four blood cells that happened to be arranged in a square pattern when the serum dried could be measured when coadding only 128 scans (Figure 21). This suggests that it is very nearly possible to measure single living cells, a very exciting possibility for biological investigations.

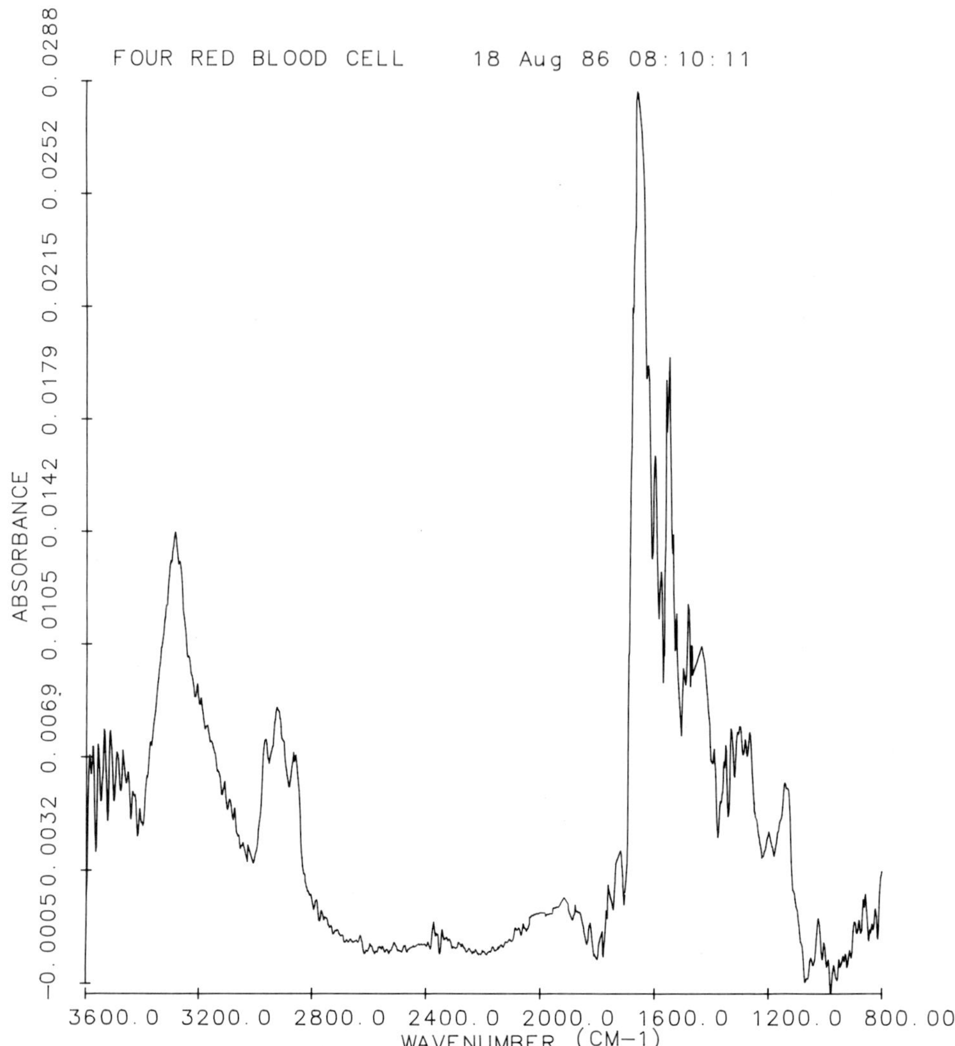

Figure 21. The infrared spectrum of four human red blood cells obtained with a total data collection time of less than two minutes (128 scans).

12.4 CONCLUSIONS

Infrared microscopy has ushered in a new dimension of problem solving capabilities to the infrared spectroscopist. It is now possible to obtain excellent data on samples which in the past have been considered difficult and unsuitable for analysis.

We feel that the infrared microscope, coupled with the sensitivity of modern Fourier transform infrared spectrophotometers, provides solutions for problems that in the past have been considered difficult, if not impossible, to solve with traditional sampling methods of infrared spectroscopy. Many of the samples discussed here could have been analyzed with various types of beam condensing accessories, albeit less conveniently. The ease of use and the benefit of being able to selectively and interactively mask off areas of interest of a sample utilizing adjustable apertures makes the infrared microscope the method of choice.

The advancements in infrared microscopy techniques will allow the investigation of ever smaller samples. This has important implications for the studies of biological samples. As it is possible to study structures on the order of 10 microns in diameter with the infrared microscope; the direct investigation of phyto-cellular substructure and chemistry will play an important role in the investigation of molecules of interest to biologists. More emphasis will be placed on the ability of infrared microscopy to solve complex structural problems which confront those investigators who are challenged with both minute amounts of sample and complex morphological problems.

REFERENCES

1. P. Roush, Ed., *The Design, Sample Handling, and Applications of Infrared Microscopes*, ASTM STP 949 (1987).
2. T. D. Dee, W. Herres, A. Simon and G. Zachman, *Microsample Analysis Using an Infrared Microscope*,in *Fourier and Computerized Infrared Spectroscopy*, SPIE Vol. 553: 228-229 (1985).
3. R. L. Barbour, M. D. Smith, D. A. C. Compton and M. Mehicic, *Nonstandard Sampling Techniques for Infrared Spectroscopy*, in *Fourier and Computerized Infrared Spectroscopy*, SPIE Vol. 553: 460-461 (1985).
4. J. G. Grasselli, M. Mehicic and J. R. Mooney, *FT-IR: Today and Tomorrow*, in *Fourier and Computerized Infrared Spectroscopy*, SPIE Vol. 553: 101-109 (1985).

5. J. C. Shearer and D. C. Peters, *The Art of FT-IR Microsampling*, in *Fourier and Computerized Infrared Spectroscopy*, SPIE Vol. 553: 285-286 (1985).
6. K. Krishnan, S. L. Hill and L. S. Gelfand, *Some Applications of Microtransmittance and Microreflectance Techniques*, in *Fourier and Computerized Infrared Spectroscopy*, SPIE Vol. 553: 338-339 (1985).
7. I. Zelicht, *Photosynthesis, Photorespiration, and Plant Productivity*, Academic Press, New York (1971).
8. A. Lee Smith, *Applied Infrared Spectroscopy, Fundamentals, Tecniques and Analytical Problem-Solving*, Wiley-Interscience (1979).
9. R. G. Messerschmidt, *Photometric Considerations on the Design and Use of Infrared Microscope Accessories*, in *The Design, Sample Handling, and Applications of Infrared Microscopes* (P. B. Roush, ed.), ASTM STP 949: 12-26 (1987).
10. I. Murray, *Near Infrared Reflectance Analysis of Forages*, in *Recent Adv. Anim, Nutr.*,141-156 (1986).
11. T. Avella, A. Copin, R. Caussin, C. Duculot and P. H. Martens, *Study of Deposits of Photopharmaceutical Products by Attenuated Total Reflectance Infrared Spectrometry*, in *Bull. Rech. Agron. Gembloux*, 5J: 341-351 (1970).
12. R. E. Baier, *Noninvasive Rapid Characterization of Human Skin Chemistry in situ*, in *J. Soc. Cosmet. Chem.*, 29: 283-290 (1975).
13. R. M. Gendreau, R. I. Lieninger and R. J. Jakobsen, *Molecular Level Studies of Blood Protein-Materials Interactions*, in *Biomaterials 1980* (G. D. Winter, D. F. Gibbons and G. Plenk, Jr., eds.), John Wiley and Sons, Ltd., pp. 415-421 (1982).
14. J. L. Koenig and D. L. Taub, *Infrared Spectra of Globular Proteins in Aqueous Solution, Analytical Applications of FT-IR of Globular Proteins in Aqueous Solution*, in *Analytical Applications of FT-IR to Molecular and Biological Systems* (J. R. Durig, ed.), D. Reidel, Dordrecht, pp. 241-255 (1980).

13

FT-IR Microspectrometry: Applications in Pharmaceutical Research

JOHN A. REFFNER *Spectra-Tech Inc., Stamford, Connecticut*

13.1 INTRODUCTION

Chemical microscopy and infrared spectroscopy are established techniques in pharmaceutical research. Now, Fourier transform spectrometry and advances in optics have made it practical to combine these sciences into a new analytical technology [1-4]. With FT-IR microspectrometry it is possible to obtain infrared spectra of microscopic samples. Microscopy and spectroscopy have common origins in optics and their historical contributions to science. Each discipline has its unique capabilities, and when united they create a new analytical methodology.

The terms describing the combined technologies are often confused with the terms used to define other infrared microsampling techniques. FT-IR microspectroscopy is the science created by combining microscopy with spectroscopy. The prefix "micro" comes from the Greek mikros, meaning small. FT-IR microsampling [5] describes infrared methods that use beam condensers or small apertures to define a sample volume without directly viewing the sample. FT-IR microspectrometry specifically defines the operation of using a microscope to see and to select the sample from which an infrared spectrum is obtained.

13.2 EXPERIMENTAL

Thermomicroscopy system: a polarized light microscope, Olympus BHA Pol or equivalent, fitted with a model FP-Mettler Instrument Corporation heating stage. The temperature scale of the instrument was calibrated using melting point standards according to the manufacturer's specifications.

Tandem FT-IR/thermomicroscopy instrumentation: a Digilab FTS-15E with microsampling accessory. This accessory was modified to accept the Mettler FP-5 micro heating stage. All glass windows of the FP-5 and the cooling fan were removed. Temperature calibration with melting point standards was necessary.

DSC equipment: Model TA-3300 DSC by Mettler Instrument Company. The temperature and heat flow scales were calibrated according to the manufacturers' recommended procedures. Samples were hermetically sealed in a dry nitrogen atmosphere. A dry nitrogen atmosphere purged the system during heating and cooling.

13.3 APPLICATIONS

13.3.1 Contaminants

The obvious applications of infrared microspectrometry in pharmaceutical research and production are identification of drug constituents and contaminants. With the microscope, individual constituents can often be distinguished by their morphology or optical properties, physically isolated and their spectrum recorded. The more common contaminants found in pharmaceuticals are artifacts that enter during processing or packaging, by-products of chemical processing and decomposition products. Also, precipitates may form while a drug is sitting on the shelf.

Contaminants can be identified by microscopical examination of their morphological characteristics, but when unique morphology is lacking spectral data becomes critically important. The microscopic shapes and forms of contaminants are important features in their analysis and tracing their origins. Such direct observations provide answers to many questions; Is the sample single phase? solid or liquid ? crystalline or amorphous? fibrous or granular? ground or precipitated? Figure 1 is a micrograph of household dust, a classic contaminant. While the microscopist may identify cat hairs, wool, pollen, wood fibers and many minerals in this sample, confirming these analyses with IR spectra is a major advance in chemical microscopy.

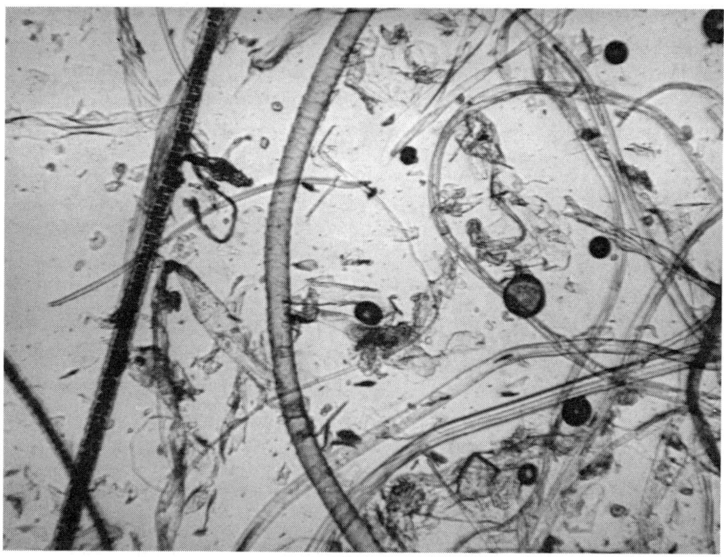

Figure 1. Photomicrograph of household dust, a universal contaminant.

13.3.2 Trace Analysis

Trace analysis is also possible with the combination of visual microscopy and FT-IR spectroscopy. Trace analyses can be divided into two types depending on how the trace component is dispersed in the major phase. Type I is the analysis of a small quantity (less than 1%) of a material uniformly distributed in a continuous phase, while Type II is a small amount of a second phase inhomogeneously dispersed in the major phase. Only Type II trace analyses are possible by infrared microspectrometry. The detection limit of FT-IR microspectrometry is demonstrated by the following example calculation: infrared spectra are readily obtained from a sample 20 μm (micrometers) in diameter and 10 μm thick. Assuming that this sample is of unit density its weight is approximately 3 ng (nanograms). If infrared spectrometry can detect ten percent of a component in a mixture, then 300 pg (picograms) of material can be detected in this sample by FT-IR microspectrometry. Certainly this level of detection qualifies as trace analysis.

A common trace analysis problem in pharmaceutical research is known as the "swirl problem". If a bottle containing a solution of a drug is picked up and swirled, and a wisp of material is seen in the vortex, then you have a "swirl problem". It's a challenge to identify these trace contaminants. Spectral analysis is essential to identifying swirls and tracing their source. With the microscope as the input device to the FT-IR spectrophotometer, high-quality infrared spectra can be obtained from the small samples seen with the microscope.

13.3.3 Polymorphism

Polymorphic behavior is a major pharmaceutical research application for FT-IR microspectrometry. Polymorphism is the existence of a substance in different forms. While a polymorphic substance has the same chemical composition in each form, the solid-state, or its liquid crystal structure is different in each form. In the solid-state, a substance may exist in more than one form. While only a single form is thermodynamically stable at any temperature and pressure, different polymorphs of a substance can be kinetically stable and exist at the same temperature and pressure. Figure 2 is a micrograph of piperamide maleate, a substance having two solid crystal forms and a liquid crystal state. The same chemical substance but different polymorphic forms.

Figure 2. Polarized light micrograph of piperamide maleate taken after 50 minutes after melt was cooled to room temperature. Three phases are shown. A and B are two solid-state forms and C is a liquid crystal phase.

Sulfur is known to exist in a monoclinic and a rhombic form. This transformation is shown in Figure 3. In this example, the rhombic sulfur form is growing while the monoclinic form is dissolving. The rhombic form was the thermodynamically stable sulfur polymorphic form at the temperature and conditions at the time this micrograph was recorded. The monoclinic sulfur phase had crystallized at a higher temperature where it was the stable phase. On cooling into the temperature domain where the rhombic form is the stable phase, there's a transformation from one form to another. This is a simple example of polymorphism. Polymorphism is an important aspect of pharmaceutical research because a drug's properties, including its bioavailability, can be different for different polymorphic forms.

Polymorphic transformations can be characterized by changes in the material's absorption spectrum. The transformation of mercuric iodine is a visual illustration of this fact, Figure 4. In transmitted light, red and yellow forms of HgI_2 are seen.

FT-IR microspectrometry has several advantages for studying the polymorphic behavior of pharmaceuticals over the traditional methods of x-ray diffraction, microscopy and thermal analysis (DTA,DSC). While conventional infrared spectroscopy has been used to identify polymorphs, these infrared

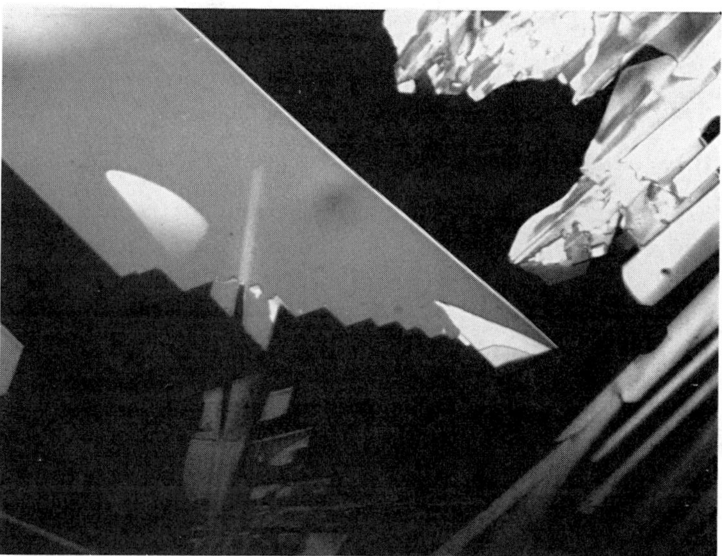

Figure 3. The polymorphic transformation of monoclinic(M) to rhombic(R) sulfur.

techniques require bulk samples of pure polymorphic forms and great care in sample preparation. Organic compounds of pharmaceutical interest are often large complex molecules and x-ray diffraction patterns and optical crystallographic properties of polymorphic forms can be very similar. Subtle differences may go undetected. In FT-IR microspectrometry, direct microscopical examination is used to isolate phases and infrared spectroscopy can be used to determine the molecular structure.

13.3.4 Polymorphism of Hexadecylaminobenboic Acid

A study of hexadecylaminobenzoic acid's (HABA) polymorphic phase behavior is presented to illustrate the synergism arising from the union of microscopy with FT-IR spectroscopy. HABA is an intermediate in the production of an experimental drug and its chemical structure is shown in Figure 5. Its polymorphic behavior is both complex and interesting.

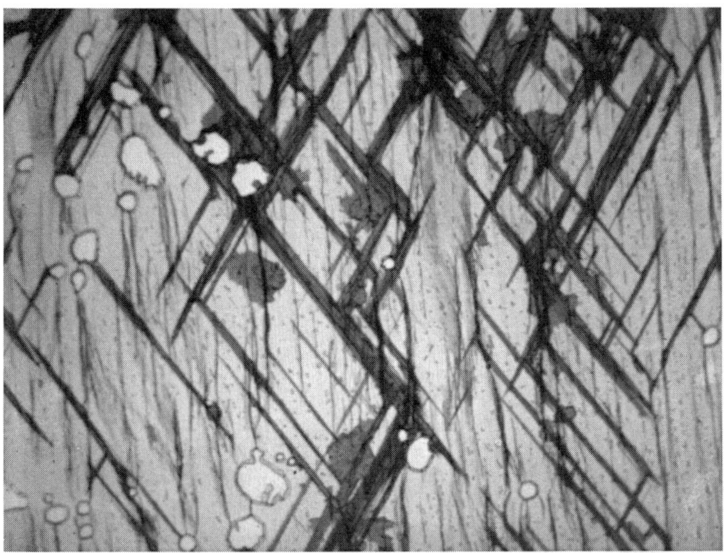

(a)

Figure 4. The solid state transformation of HgI_2 from the yellow (light) to the red (dark) polymorphic forms.

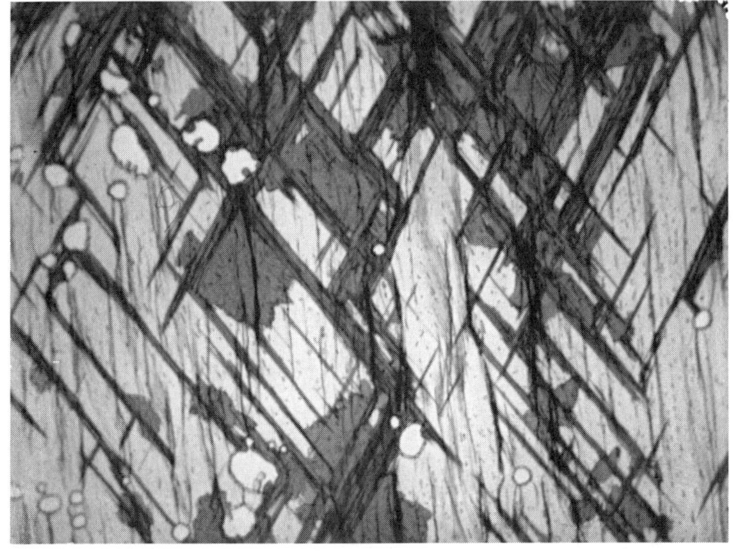

(b)

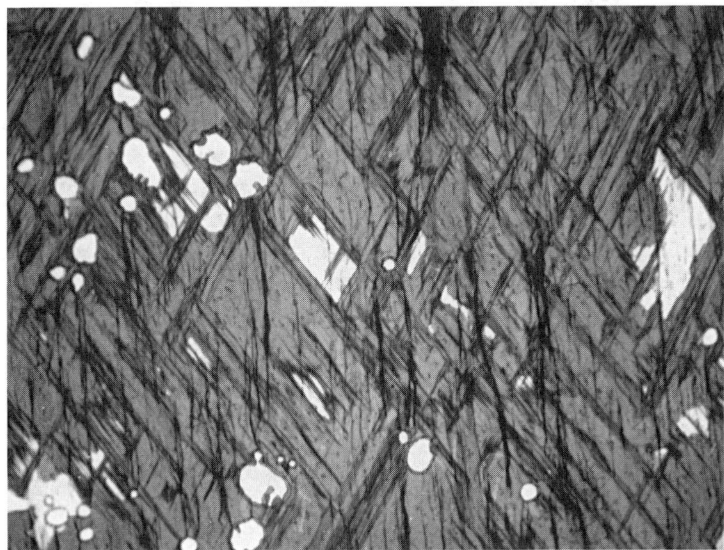

(c)

Figure 4. (continued)

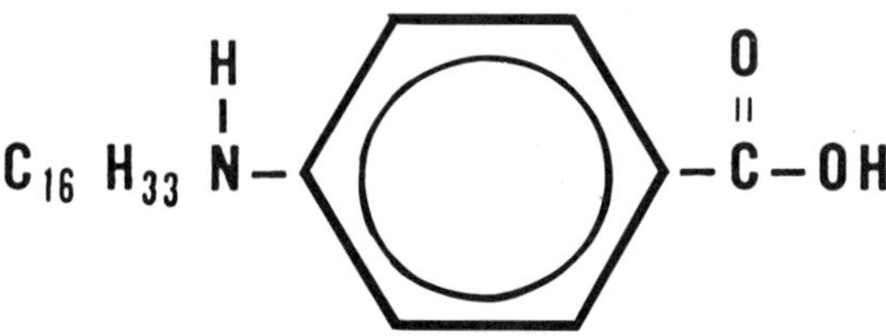

Figure 5. Structural formula of hexadecylaminobenzoic acid.

In order to perform the experiments, a special sample holder was made for the hot stage. KCl plates were fixed to the end of a 2.5 x 0.625 x 0.037 inch steel plate. The KCl plates were cleaved from single crystals and were about 1 mm thick plates, 0.5 x 1.0 cm on edge. Small samples of HABA were placed in the hot stage, observed visually and infrared absorption spectra recorded at various temperatures. One reason for combining microscopy and infrared spectroscopy is the ease of sample handling and sample preparation. Hot stage microscopy is well-established [3] and combining this with IR microspectrometry extends the analytical capabilites of both techniques [4].

The study of HABA started with differential thermoanalysis not with microscopy or spectroscopy. When HABA was heated in a DSC cell, a series of transformations were observed. Polymorphism was suspected but could not be defined. During the first heating cycle, endothermic transformations occured at 93°, 107°, and 127°C. After cooling and recrystallizing, this sample is reheated and a different pattern of transformation temperatures is recorded by DSC. Before going into the analytical details, the final phase equilibrium phase diagram is presented to aid in our discussion, see Figure 6. The room temperature stable phase is defined as Form I. When heated initially, this material transforms to a different solid-state phase, Form III at 93°C. At 105°C Form III melts to a liquid crystal state. The liquid crystal state persists up to 124°C, where it melts to the isotropic liquid. Upon cooling the phase behavior is only reversible down to 93°C. The liquid crystal phase forms from the melt at 124°C and Form III crystallizes at 105°C. However, Form III does not change at 93°C. Form III remains unchanged on cooling to 74°C, where it transforms to Form II. Once Form II has formed it remains the solid phase, even when it is cooled to room temperature. The Form II to Form I transformation is monotropic. Form II will only transform to Form I when a

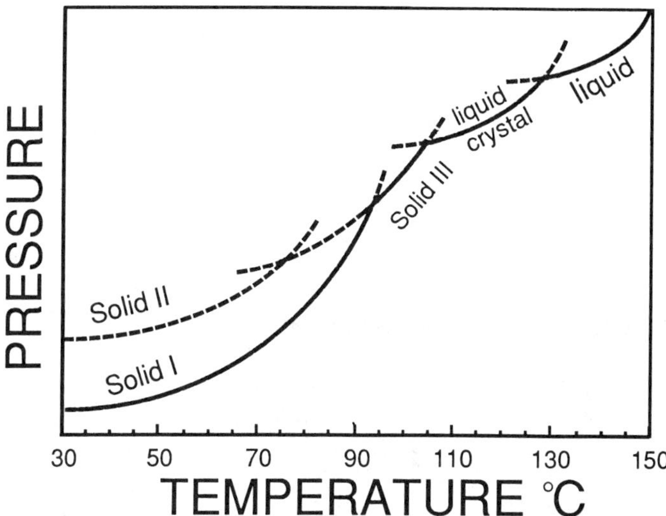

Figure 6. The phase diagram for HABA.

solvent (i.e. toluene) is present. When Form II is heated, the transformation to Form III is observed at 74°C.

HABA's numerous phases are not its only distinctive behavior, its solid state transformations are difficult or impossible to observe by light microscopy and x-ray diffraction. I know several light microscopists (I include myself among them), who insist that you can always detect polymorphic transformations by polarized light microscopy. When one solid-state form transforms to another solid-state form, the change generally can be seen, but not with HABA. Figure 7 illustrates what is seen on heating HABA through its Form I to III transformation. While there is no geometrical change in these crystals, their contrast changes. This micrograph was taken using crossed polarization, the change in contrast is actually a change in the optical retardation of these crystals during the transformation. This is a subtle change and the experiment had to be repeated several times to verify that a transformation occurred. Without the prior DSC data to indicate that a thermally induced change was occurring at this temperature, this transformation would not have been seen. The transformation of HABA's Form II to III is unobservable by polarized light microscopy. Figure 8 shows the same field

Figure 7. Polarized light micrographs showing the transformation of HABA's Form I to Form III.

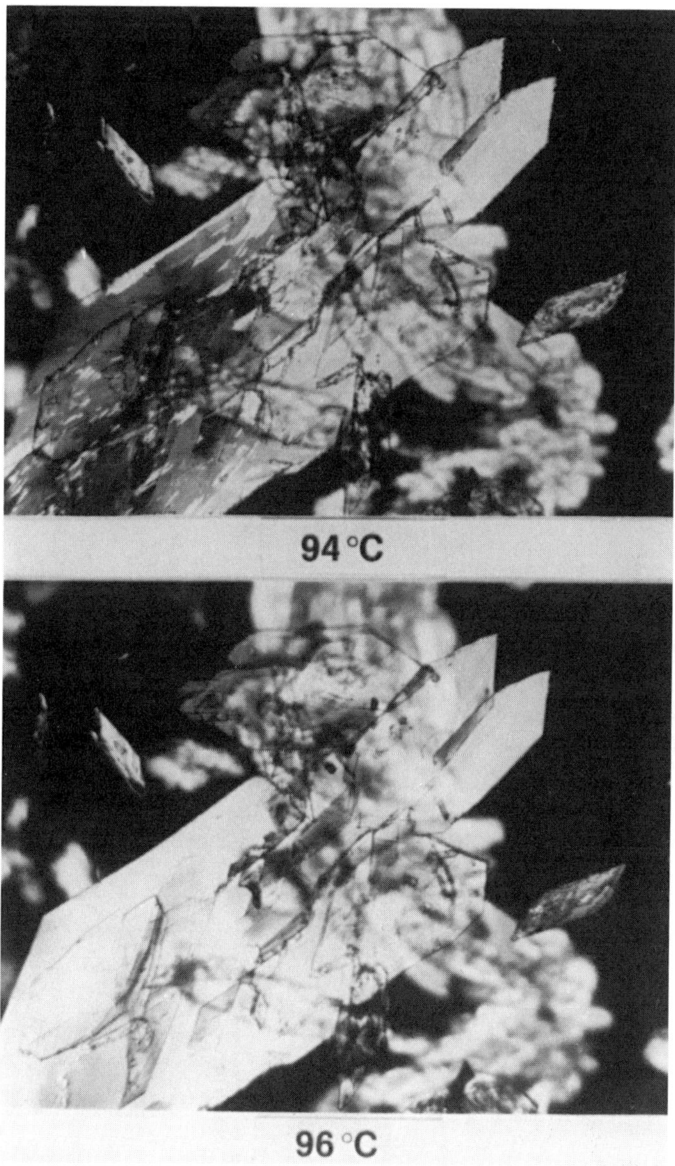

Figure 7. (continued)

Figure 8. Polarized light micrographs of the same field before and after the Form II to Form III solid-state transformation of HABA.

FORM III 83 °C

Figure 8. (continued)

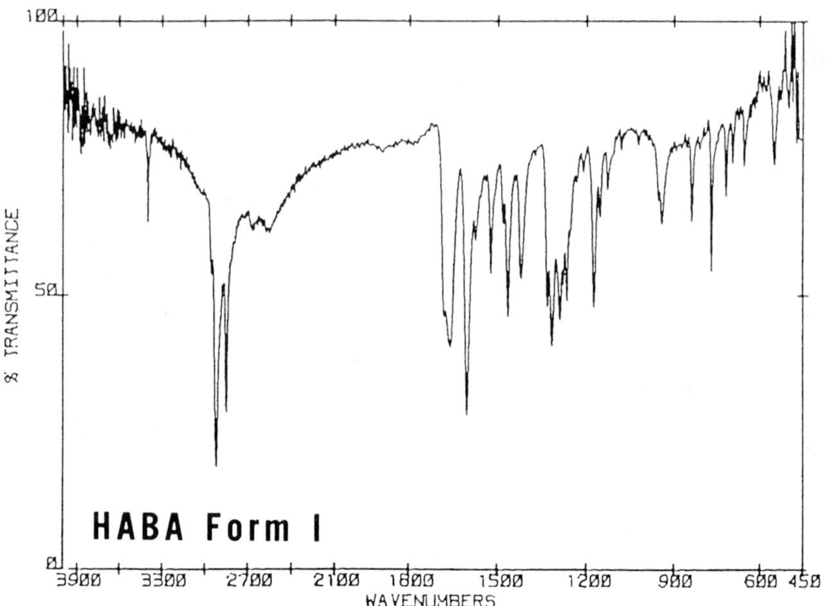

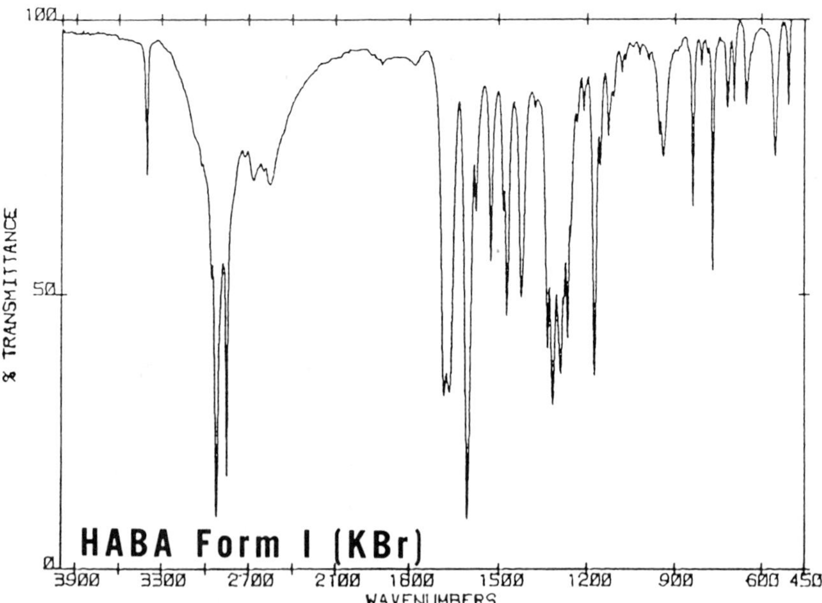

Figure 9. Infrared absorption spectra of HABA solid-state phases, recorded by both microspectrometry and conventional KBr techniques.

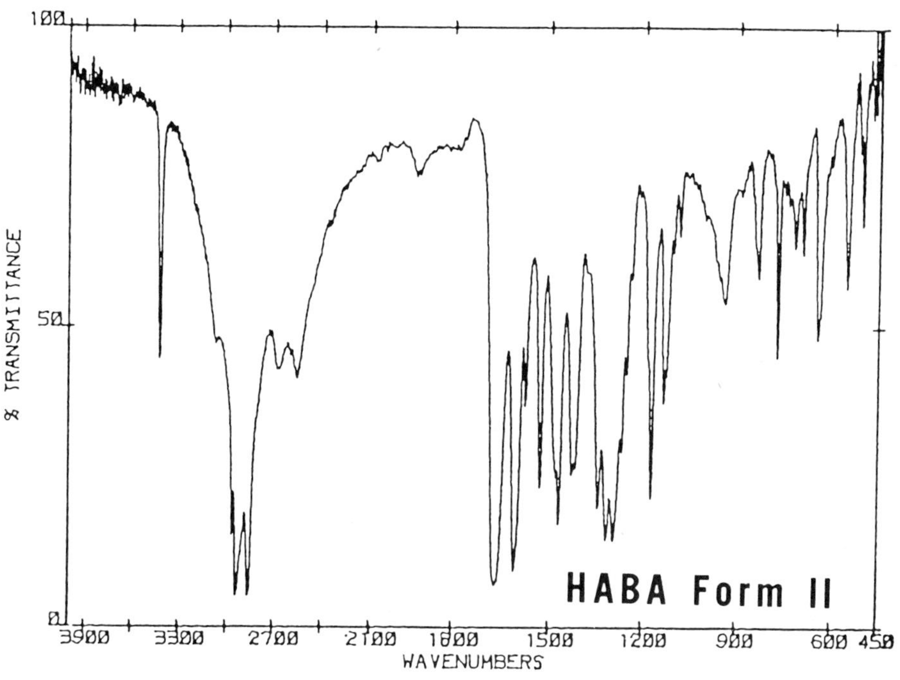

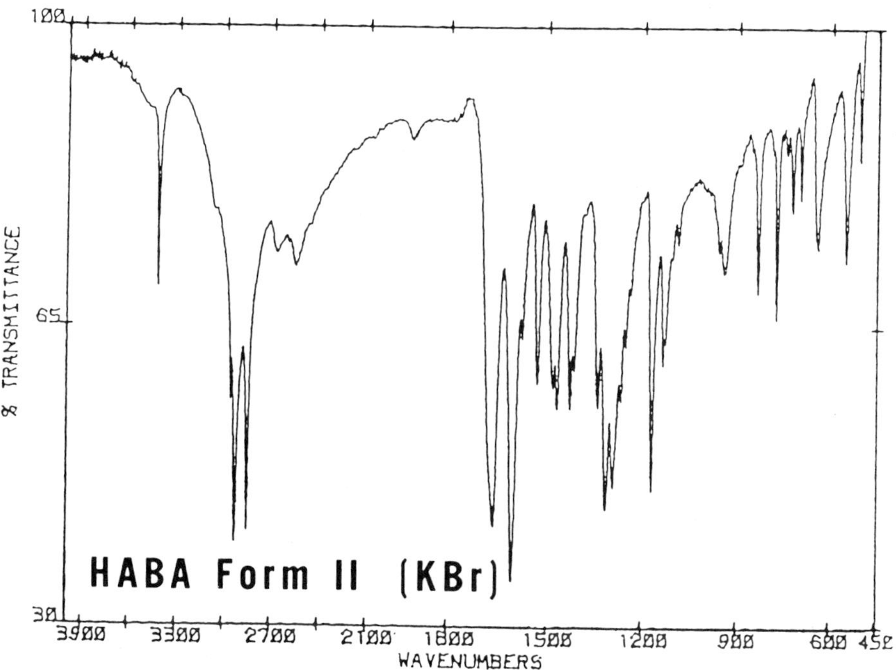

Figure 9. (continued)

before and after the Form II to III transformation. No visual change was seen for this transformation.

Infrared spectroscopy can distinguish between a chemical reaction and a polymorphic transformation. Polymorphic changes neither result from, nor do they cause changes in a substance's primary molecular bonding. In the study of polymorphism, infrared spectroscopy provides answers to two questions; did a change occur and if a change occurred did the primary molecular bonding remain unchanged? A substance's infrared spectrum is a record of its molecular bonding. While minor changes in infrared spectra are found for different polymorphic forms, breaking or making of new primary chemical bonds indicates new chemical structures formed. Comparing the infrared spectra of different phases readily distinguishes chemical changes from polymorphic transformations.

While microscopy has been a primary technique for studying polymorphism, FT-IR microspectrometry expands its capabilities. Again HABA is an example. HABA's Form I and II can be isolated at room temperature and their infrared spectra can be obtained by either microscopic or conventional KBr sampling techniques. Comparing spectra of the same forms recorded by these techniques, Figure 9(a) with 9(b) and 9(c) with 9(d), only minor differences are seen in the relative band intensities. Preferred orientation of small microscopic samples may account for these variations. When the infrared spectra of the two polymorphic forms are compared, Figure 9(a) with 9(c) for example, their similarity demonstrates that the primary molecular bonding is unchanged, confirming their polymorphic relationship.

Adding a heating stage to the FT-IR microscope extends the study of HABA's phase behavior to elevated temperatures. Figure 10 reports the IR spectra for all HABA phases in the region from 1800 to 900 cm^{-1}. While all spectra have similar absorption bands, each phase has distinct features.

The DSC technique detects thermal transitions without knowledge of what changes are taking place. DSC is blind. Alone, DSC cannot know that a liquid crystal phase formed, that a crystalline phase melted or that a polymorphic transformation took place. Direct microscopical examination is necessary to confirm the nature of these changes. The added power of FT-IR microspectrometry makes it possible to examine the changes in molecular bonding accompanying the changes seen with the microscope.

HABA's infrared spectrum can be interpreted by using group frequency correlations, but each phase exhibits minor spectral variations. These differences distinguish each phase. The N-H band is present in the spectra of all phases, in Form I it is found at 3400 cm^{-1}, while in all other phases it is at 3420 cm^{-1}. Form I also has a doublet at 1660 and 1682 cm^{-1}, but this acid carbonyl band is a singlet occuring at 1665 cm^{-1} for Form II, at 1670 cm^{-1} for

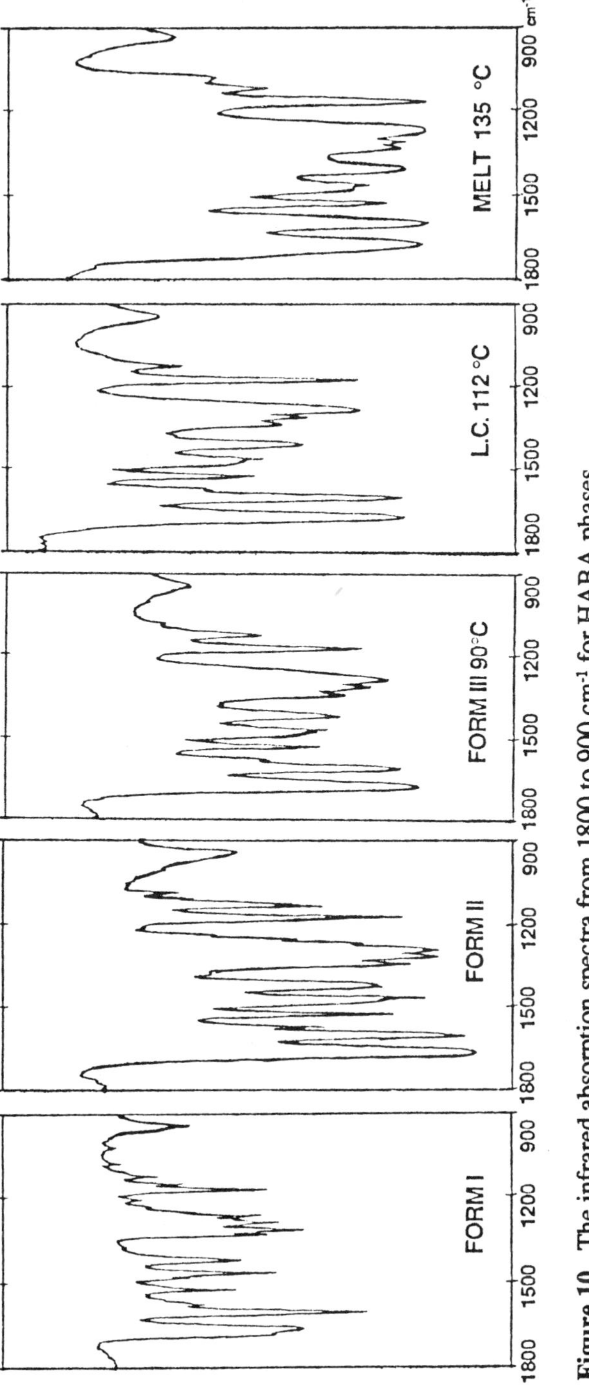

Figure 10. The infrared absorption spectra from 1800 to 900 cm^{-1} for HABA phases.

Form III, at 1673 cm^{-1} for the liquid crystal, and1677 cm^{-1} for the isotropic molten liquid. These spectral changes result from the geometrical packing of the molecules and changes in the molecular bonding. In Form I the higher energy N-H stretch and carbonyl suggest that the aromatic ring delocalization is weaker in this form. The most probable cause for this would be the loss of planarity of the hydrogen bonded dimer of the carboxylic groups with the aromatic ring. The differences between all other phases result in the packing of the aliphatic hydrocarbon chain.

13.4 CONCLUSION

Infrared microspectrometry is a valuable analytical tool for the pharmaceutical chemists. The polymorphic behavior of HABA was chosen to demonstrate how the combination of the microscope and the infrared spectrometer can join with other technologies to resolve a difficult analytical problem. Infrared spectroscopy and microscopy are fundamental scientific disciplines, by their combination a new technology has evolved that increases our ability to solve complex analytical problems.

REFERENCES

1. R. G. Messerschmidt, J. M. Kwiatkoski and C. R. Friedman, *International Labmate*, XI: 12-16 (1986).
2. H. J. Humecki, and R. Z. Muggli, *Micro sample identification with FTIR spectroscopy*, in *Microbeam Anal.*, 17: 243-246 (1982).
3. K. Krishnan, *Applications of FT-IR microsampling techniques to some polymer systems.* in *Polym. Prepr. (Am. Chem. Soc., Div. Polym. Chem.)*, 25(2): 182-184 (1984).
4. J. N. Ramsey, and H. H. Hausdorff, *Applications of small-area infrared analysis to semiconductor processing problems.* in *Microbeam Anal.*, 16: 91-95 (1981).
5. R. Cournoyer, J. C. Shearer, and D. H. Anderson, *Anal. Chem.*, 49: 2275 (1977).
6. M. Kuhnert-Brandstatter, and S. Wunsch, *Mikrochim. Acta* (Wein), 1297 (1969).

Miscellaneous Applications of Infrared Microspectroscopy

14

HPLC/FT-IR Measurements by Transmission, Reflection-Absorption, and Diffuse Reflection Microscopy

DAVID J. J. FRASER, KELLY L. NORTON, AND PETER R. GRIFFITHS *Department of Chemistry, University of California, Riverside, California*

14.1 INTRODUCTION

Most applications of FT-IR microscopy which have been described to date have involved measuring the spectra of samples which are of sufficient thickness (5-50 μm) to yield an adequate infrared spectrum when at least one of the other dimensions is smaller than the typical beam diameter at the focus of a 6x beam condenser (1-2 mm). These measurements involve little in the way of sample preparation except immobilizing the specimen on the surface of a suitable substrate, possibly after microtoming to obtain the correct sample thickness. The substrate is then mounted on the X-Y-Z stage of the microscope and moved until the region of the sample which is of interest is located at the beam focus.

For a beam diameter of 50 μm and a sample thickness of 5 μm, the weight of analyte being observed is about 10 ng, assuming a density of 1 g mL^{-1}. From other reports in this volume, it can be seen that a sample of these dimensions can yield a spectrum of high signal-to-noise ratio (SNR) in a fairly short measurement time. FT-IR microscopy therefore represents a potentially powerful means of measuring the infrared spectrum of any analyte in low nanogram and even high picogram quantities, provided that the sample can be contained in a small area.

We have applied this principle to the measurement of the infrared spectra of compounds eluting from capillary gas and supercritical fluid chromatographs [1-6]. In both cases, the column effluent is passed through a narrow fused silica tube held just above a moving ZnSe window. For capillary gas chromatography (GC), the internal diameter (i.d.) of this tube is about 50 μm, which is too large to reduce the flow rate of the carrier gas significantly. For capillary supercritical fluid chromatography (SFC), the i.d. of the end of the tube is about 1 μm, so that in this case the tube acts as a restrictor to maintain the pressure in the column at the desired level. The ZnSe window is cooled below ambient temperature, so that each eluting component is trapped on the plate while the mobile phase remains as a vapor and is not condensed. If the distance between the end of the fused silica tube and the ZnSe window is between about 50 and 100 μm, each component may be trapped as a very small spot.

If the window is held stationary, the profile of the thickness of a GC peak trapped in this manner is approximately Gaussian [1], with a full-width at half-height (FWHH) of about 100 μm. Because of the small diameter of the trapped analyte, it is obviously necessary to measure its infrared spectrum using an FT-IR microscope.

To understand the benefits to be gained by trapping each component in as small an area as possible, let us first consider the relative thickness of 10 ng of a certain component deposited as a spot with (a) a FWHH of 100 μm, and (b) a FWHH of 1 mm. For simplicity, let us assume that the sample is deposited as a cylinder whose diameter is equal to the FWHH of the Gaussian deposit, and let us again assume a density of 1 g mL^{-1}. The thickness of the 100 μm diameter spot is 1.3 μm, which is certainly sufficient to yield an adequate spectrum of any organic compound with a minimal amount of ordinate scale expansion. For most polar compounds, the strongest band in the spectrum of a sample of this thickness would be expected to have an absorbance of between 0.2 (63% transmittance) and 0.1 (80% T). The thickness of the 1 mm diameter spot is 0.013 μm, so that the absorbance of the strongest band would be expected to be between 0.002 (99.5% T) and 0.001 (99.8% T). Obviously in this case the amount of ordinate expansion required to visualize the spectrum is very great and requires an exceptionally low spectral noise level, remarkable baseline flatness and essentially no interference from atmospheric (or any other) species.

If we assume that the beam diameter can be controlled so that it is always equal to the diameter of the deposit, the area of the beam interrogating the 1 mm spot is one hundred times greater than that interrogating the 100 μm diameter spot. The signal at the detector should therefore be one hundred times greater in this case, so that (all things being equal) the baseline noise level in a ratio-recorded spectrum of the 1 mm spot would be two orders of magnitude lower than that for the 100 μm spot. All things are not equal, however [6]. First, the SNR at the centerburst of an interferogram of a 1 mm sample held at the focus of a standard 4x or 6x beam condenser and measured using a narrow range or medium range mercury cadmium telluride (MCT) detector is limited by the dynamic range of the analog-to-digital converter. The high intensity of the beam will also usually drive the detector or its preamplifier nonlinear. To rectify both problems, the beam must be attenuated in some manner between the source and the detector when the beam diameter at the focus of an FT-IR microscope exceeds about 250 μm.

In addition, it should be noted that a smaller detector can be used for measuring the interferogram of the smaller sample. Since the noise equivalent power of a detector with a square element is proportional to its linear dimension, a further benefit of between four and ten may be obtained in this way. Finally the extent of beam condensation may be increased above that of a 6x beam condenser through the use of an infrared microscope, giving rise to an improvement of about a factor of two. Because of the number of mirrors involved, infrared microscopes are not quite as efficient as beam condensers, which slightly offsets the advantages listed above.

In summary, therefore, there is a benefit of anywhere from a factor of ten to almost one hundred to be gained in GC/FT-IR and SFC/FT-IR measurements by depositing each eluting component as a spot of about 100 μm diameter and measuring the spectrum with an FT-IR microscope in comparison to the case where the sample is deposited as a 1 mm diameter spot and the spectrum is measured using a conventional beam condenser.

We have been attempting to apply similar principles to the measurement of the infrared spectra of components eluting from a high performance liquid chromatography (HPLC) column. Several techniques for measuring HPLC/FT-IR spectra have been reported, the two most common of which are deposition on a moving infrared-transparent window, followed by transmission spectrometry [7-10], and depositing the material on an alkali halide powder held in a small cup, followed by diffuse reflection (DR) spectrometry [11-16].

The first technique, which has been termed the "buffer memory" technique by Jinno, has the advantage over the DR method that it is considerably simpler and allows a continuous deposition so that chromatograms can be reconstructed from the infrared spectra. Its disadvantages include the fact that:

(a) The technique is only readily practicable with columns of very small diameter, as it is difficult to eliminate more than five microliters of solvent per minute in the manner used by Jinno. Thus packed fused silica columns with internal diameters of 500 μm or less (which are not commercially available in the U.S.) must be used, and the capacity of these columns is quite low (of the order of 1 μg per component).

(b) The percent absorption of the incident radiation by a particular band is smaller when measured by transmission than when measured by diffuse reflectance.

(c) The diameter of the spot occupied by each deposited component in the buffer memory method is rarely less than 1 mm, so that the advantages of FT-IR microscopy described above have not yet been realized in practice in any HPLC/FT-IR interface.

The diffuse reflectance technique has certain advantages over the buffer memory method, the most important of which has usually been considered to be the capability of eliminating greater volumes of solvent in a given time because of the greater specific surface of the powdered substrate. In addition, as noted above, DR spectrometry is a very sensitive method of measuring the infrared spectrum when less than 1 µg of sample is available. However, several disadvantages serve to offset these apparent benefits, including the following:

(a) The powdered substrate must be held in discrete cups to prevent the sample spreading over an excessively large area.

(b) If more than one drop of the column eluate reaches a given cup, the solute which had been deposited on the uppermost layers of the powder may often be washed deeper into the cup and may not be observed by the diffusely reflected beam [13-15].

(c) The system is mechanically complex, requiring the sample cups to be advanced to exactly the same position in the infrared beam after each drop has fallen.

(d) Refilling every cup at the completion of the chromatographic separation can be quite time-consuming.

(e) Because the solute present in each drop is deposited on a different cup, the chromatogram must be reconstructed by interpolating data points which are often quite widely spaced.

It is apparent that a technique encompassing the strong points of the buffer memory and diffuse reflectance techniques is desirable, where the disadvantages associated with each technique are minimized. The system should be able to be operated with commercially available HPLC columns - preferably of 4.6 mm, but at least 1.0 mm, internal diameter. Solutes should be deposited continuously in as small an area as possible, and chromatographic resolution should obviously be maintained. The infrared sampling technique should yield spectra of the greatest possible SNR, with detection limits in the low nanogram (and preferably high picogram) range.

In this chapter, we will describe a device by which components eluting from a normal phase HPLC column can be deposited in spots of approximately 100 µm diameter. Several substrates and sampling techniques were investigated. For solutes deposited on a ZnSe window, transmission spectra were measured in a similar manner to the GC/FT-IR and SFC/FT-IR interfaces developed in our laboratory [1-6]. For samples deposited on a flat metallic substrate reflection-absorption (R-A) spectra were obtained in an analogous fashion to the HPLC/FT-IR interface reported recently by Gagel and Biemann [17]. Finally, samples were deposited on strips of powdered KCl in a similar manner to the SFC/FT-IR interface described by Shafer *et al.*[3].

14.1.1 Solvent Elimination Devices

We have tested several designs for devices to cause solvents emerging from 1 mm i.d. HPLC columns at a rate of about 50 μL min^{-1} to be completely evaporated while solutes are deposited as very small spots on a suitable substrate. The two most successful of these are illustrated schematically in Figure 1. The first simply involved passing the effluent from the column through a 50 μm i.d. fused silica tube, around the end of which is wrapped a short length of nichrome heating wire, see Figure 1A.

For the second, the effluent is also passed through a 50 μm i.d. fused silica tube (of 150 μm outside diameter). This tube is held inside a 500 μm i.d. fused silica tube, which is heated in a similar fashion to the device shown in Figure 1A. The inner tube extends about 2 mm from the end of the outer tube, see Figure 1B. Nitrogen gas is passed through the 500 μm i.d. tube. The hot gas heats the inner tube so that the temperature of the solution is raised to the point that a small fraction of the solvent is vaporized, causing the remaining solution to be nebulized. The droplets of the solution emerging from the inner tube are entrained as a narrow stream by the flowing hot nitrogen, which also causes most of the remaining solvent to evaporate. If a substrate is held about 4 mm from the end of the inner tube, the solute can be deposited over a very small area provided that the last remaining solvent can be evaporated rapidly. If the rate of evaporation is slow and the substrate is stationary, any solute already on the substrate is redissolved and redeposited in a circular pattern when the solvent finally evaporates, see Figure 2A. The diameter of the circle is determined by the rate of evaporation; the slower the evaporation rate, the greater the diameter of the circle. If the substrate is translated at a constant speed, solutes are deposited as two "tracks" on either side of the point immediately under the end of the inner tube, as shown schematically in Figure 2B, causing both chromatographic resolution and infrared sensitivity to be reduced. In our work, it was found that the type of deposition illustrated in Figure 2 occurred more frequently with the solvent elimination device shown in Figure 1A than with the somewhat more complicated device shown in Figure 1B. Thus, the latter device was used for all the results described in the final section of this paper.

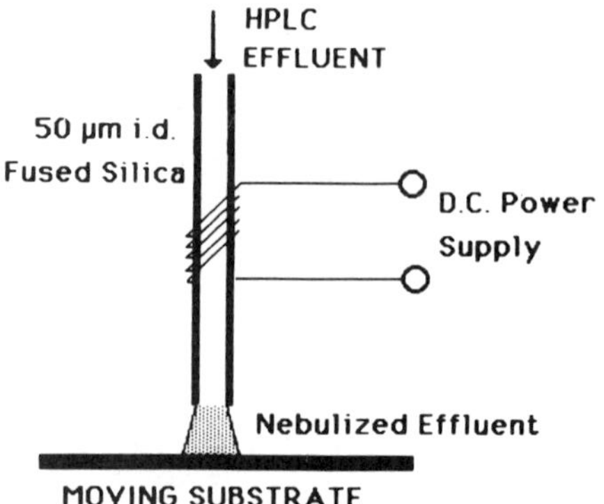

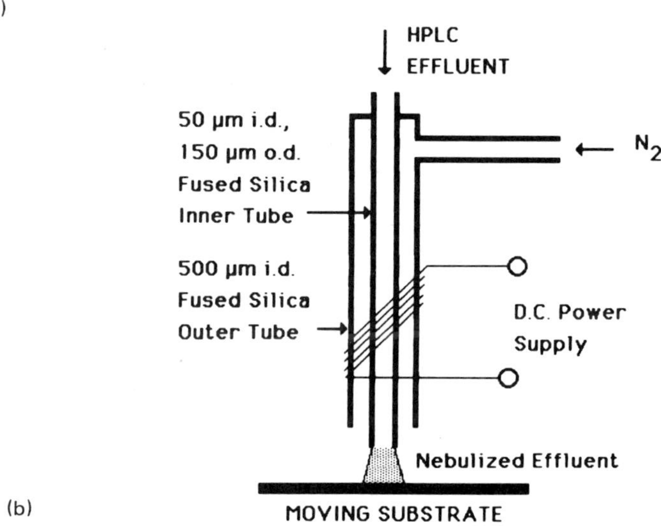

Figure 1 (a). Simple device for eliminating the solvent from effluents from 1 mm i.d. HPLC columns. **(b).** Improved device for solvent elimination from 1 mm i.d. HPLC columns.

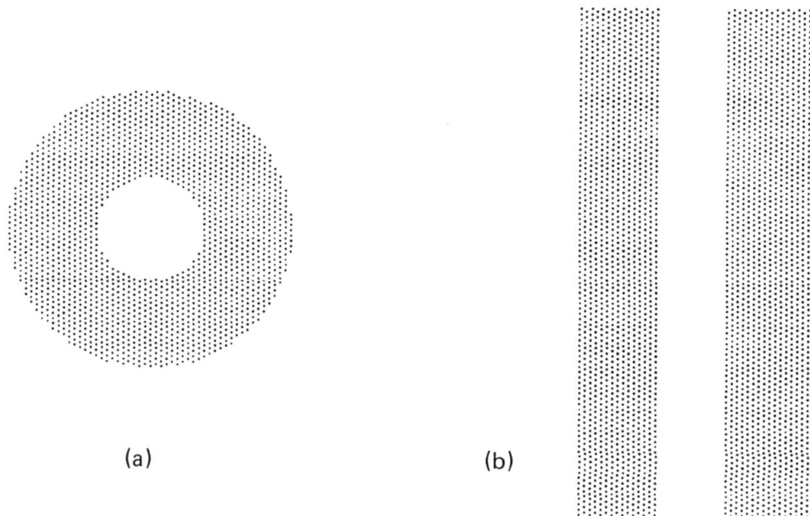

Figure 2. Schematic representation of the way in which solutes are deposited on **(a)** a stationary substrates and **(b)** a moving substrate when the solvent does not completely evaporate before reaching the substrate.

The success of this approach varied with the nature of the substrate. With flat shiny metal plates, correct depositions were particularly difficult to achieve, and deposits usually had the appearance shown schematically in Figure 2. For stationary ZnSe windows and slightly roughened metallic substrates, solutes could be deposited as spots of about 250 μm under optimized conditions. If deposition was performed on a plate on which an approximately 250-μm thick layer of KCl powder had been set down in a similar manner to the SFC/FT-IR substrate described by Shafer *et al.* [3], solvent evaporation was greatly facilitated, and solutes could be deposited as spots as small as 100 μm in diameter.

14.2 EXPERIMENTAL

All chromatographic separations were performed using a 25 cm long, 1 mm i.d. column packed with a 5 μm cyano-bonded silica stationary phase. The mobile phase was a 10:30 (v:v) mixture of $CF_2Cl\text{-}CCl_2F$ and CH_2Cl_2 at a flow rate of 50 μL min^{-1}. The volume of the injection loop was 0.5 μL.

Spectra were measured using a Perkin-Elmer (Ridgefield, CT) Model 1800 FT-IR spectrophotometer equipped with a 1 mm square, medium-range MCT detector ($D^*_{peak} = 3x10^{10}$ cm $Hz^{1/2}$ W^{-1}; λ max = 13 µm). A Spectra-Tech (Stamford, CT) IR-plan microscope was used either in the transmission or reflection mode for all measurements. After transmission through or reflection from the sample, the beam was passed back into the spectrophotometer and interferograms were measured using the detector which was installed in the spectrophotometer rather than a special-purpose detector installed in the microscope itself. All spectra were measured by signal-averaging 16 scans at 4 cm^{-1} resolution (~5 s acquisition time).

As noted above, the detector element was 1 mm square, so that the use of a smaller detector would have been beneficial. For example, a fourfold improvement in SNR should have resulted if a 0.25 mm detector with the same D* had been used for the measurement of spectra with apertures of 100 µm or less. Since the efficiency of the optics used to transfer the beam back to the spectrophotometer was on the order of 50%, a theoretical improvement in SNR of almost an order of magnitude could be predicted if an optimized detector installed in the microscope itself had been used.

A simple mixture of two quinones, phenanthrenequinone and acenaphthenequinone, at the appropriate concentrations in dichloromethane, was used for all measurements reported in the next section. Both these solutes are quite non-volatile and have band absorptivities which are typical of most polar molecules.

14.3 RESULTS AND DISCUSSION

As discussed in the Solvent Elimination Devices section, solutes were deposited on three types of substrates, a flat metal plate, a ZnSe window, and a metal strip on which a layer of KCl powder had been deposited (which we will refer to as a "KCl strip"). Initially we believed that the most promising substrate would be the flat metal plate, in view of the following:

(a) its temperature is more easily controlled at the point of impact of the drops than for any other substrate because of the high thermal conductivity of metals;

(b) studies by Gagel and Biemann [17] have indicated that solutes could be easily deposited on metals, albeit not in as small an area as we desire;

(c) we could conceive of a simple device based on a continuous "moving belt", which would be similar in principle to a moving belt HPLC/MS interface [18-20] but where each component is deposited over a much smaller area.

In practice, several difficulties were encountered when metallic substrates were tested. When the metal had a mirror finish, the sample usually spread to a

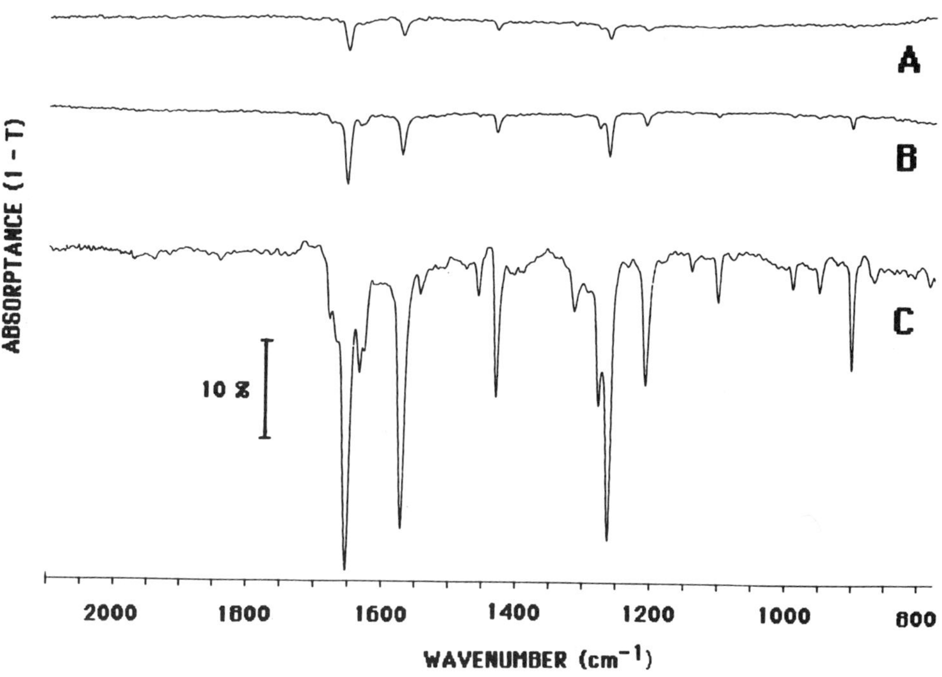

Figure 3. Spectra of 50 ng (injected) of phenanthrenequinone measured by **(a)** spectrometry after deposition on a moving plate, **(b)** transmission spectrometry after deposition on a ZnSe plate, and **(c)** DR spectrometry after deposition on a KCl strip.

diameter of at least 500 µm whether the device shown in Figures 1A or 1B was used. Better results were found using a metallic substrate with a slightly roughened (i.e., visually dull) surface, especially when the solvent was eliminated by the device shown in Figure 1B. In this case, the surface irregularities appear to restrict spreading of the solution and, possibly, to assist solvent evaporation because of the increased surface area. Spectra of samples at levels of several tens of nanograms could usually be measured with good sensitivity, albeit with some band distortion due to the effect of anomalous dispersion.

The R-A spectrum of 50 ng (injected) of acenaphthenequinone deposited on a rough aluminum substrate, measured by reflection microscopy with a 100 μm aperture, is shown in Figure 3A. 50 ng of phenanthrenequinone eluted under the identical conditions for Figure 3A were then deposited on a ZnSe window, and the spectrum was measured by transmission microscopy, see Figure 3B. It is noteworthy that the absorbance of the carbonyl band in the transmission spectrum is greater than in the R-A spectrum, despite the fact that the effective pathlength for the R-A spectrum is almost three times greater than for the transmission spectrum (since in R-A spectrometry, the beam passes through the sample twice, at an angle of about 45°).

Although this result indicates the superiority of transmission spectrometry over R-A spectrometry for HPLC/FT-IR measurements, we were still not convinced that our goal of a 1 ng detection limit could be reached with either sampling technique. In view of our previous experience with DR spectrometry applied to both HPLC/FT-IR [13-17] and SFC/FT-IR [3] interfaces, we decided

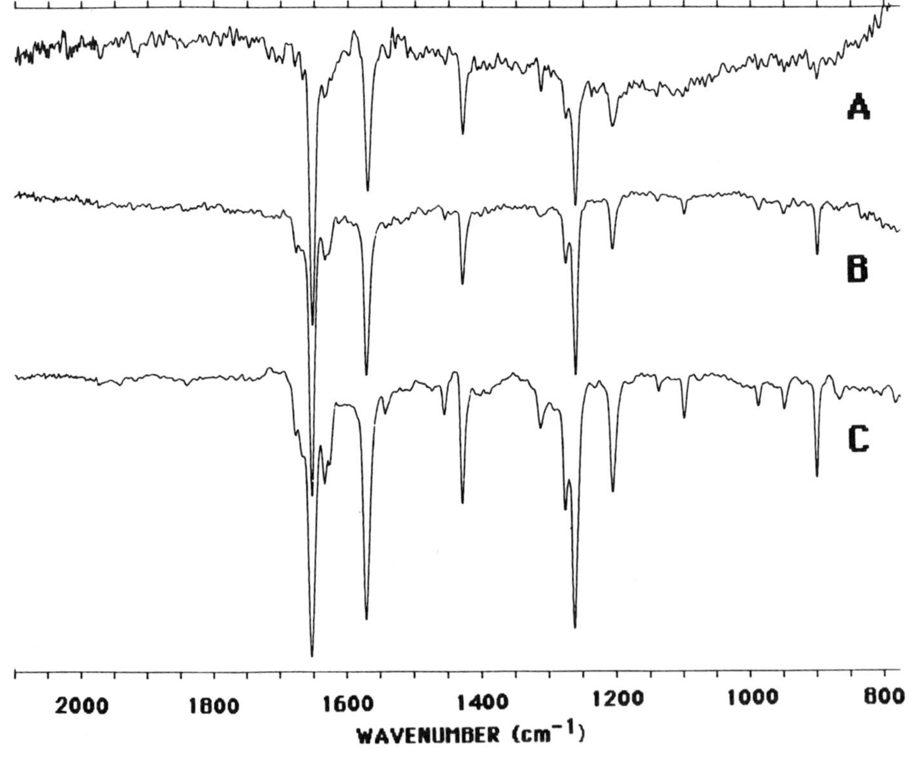

Figure 4. Spectra shown in Figure 3 after ordinate expansion so that the carbonyl band has the same amplitude in each spectrum.

to investigate the feasibility of depositing each component as a very small spot on the KCl strip and measuring the DR spectrum of the resulting spot.

For the CF_2Cl-CCl_2F:CH_2Cl_2 solvent system being used for the separation of the test mixture, it was found that the quinones could be deposited as spots of less than 100 μm diameter on the KCl strip. Their DR spectra were measured using the microscope in its "specular reflection" mode. Although the energy reaching the detector was much lower than for the transmission and R-A measurements, the increased baseline noise was more than compensated by the increased signal. Using the same chromatographic conditions as those used to obtain the transmission and R-A spectra shown in Figures 3A and 3B, the diffuse reflectance spectrum shown in Figure 3C was obtained. It is readily apparent that the intensities of all bands across the spectrum were increased. Figure 4 shows the corresponding spectra shown in Figure 3 with the intensities

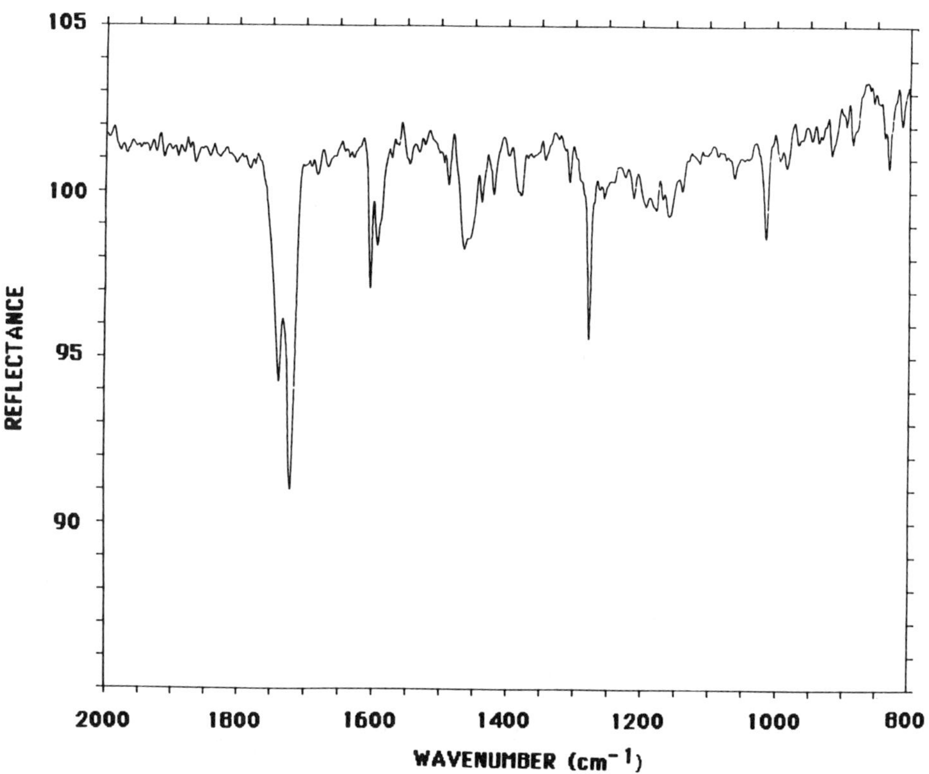

Figure 5. DR spectrum of 5 ng (injected) of acenaphthenequinone after deposition on a KCl strip.

of the carbonyl peak normalized with respect to each other. It can easily be seen that bands at longer wavelengths in the R-A and transmission spectra are diminished in intensity with respect to those of the diffuse reflectance spectrum. Furthermore, the intensities of the long wavelength bands relative to, say, the carbonyl stretching band at 1720 cm^{-1} appears to be somewhat greater in the transmission spectrum than in the R-A spectrum of the same sample quantity.

It was determined by similar experiments at the 5 ng injection level that the relative sampling efficiency of the methods and their relative sensitivity remained about the same. The spectrum of a 5 ng injection of acenapthenequinone measured by diffuse reflectance microscopy is shown in Figure 5. This spectrum would certainly permit identification of the sample by

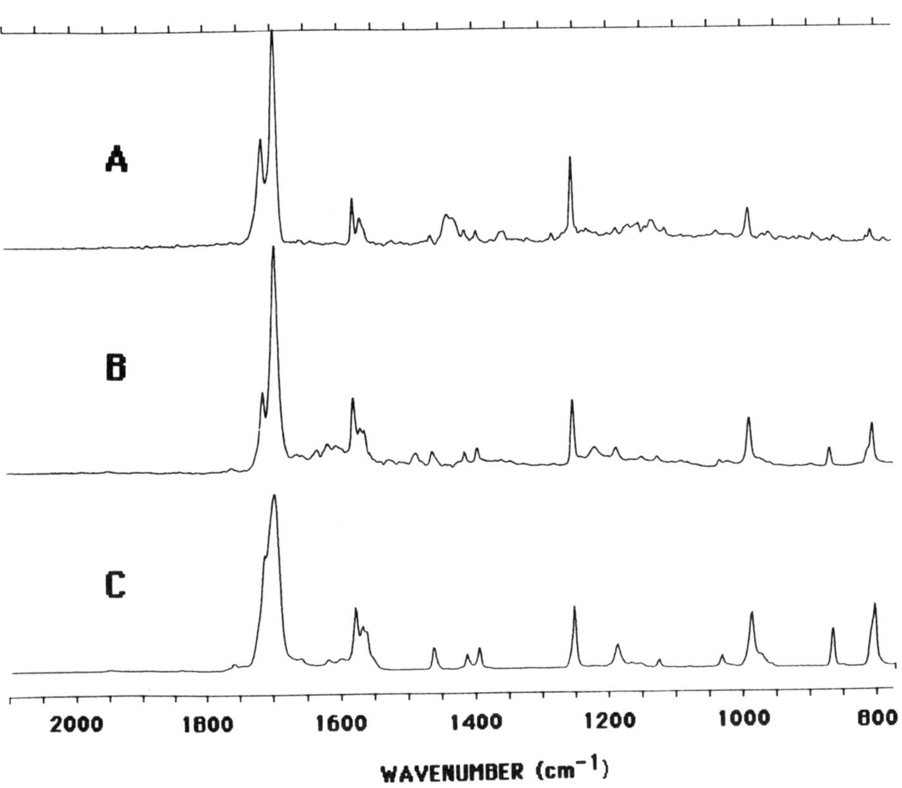

Figure 6. **(a)** Spectrum shown in Figure 5 after conversion to the Kubelka-Munk format. **(b)** Kubelka-Munk spectrum of several micrograms of acenaphthenequinone measured in a conventional DR accessory. **(c)** Linear absorbance spectrum of a KBr disk containing about 1 mg of acenaphthenequinone.

comparison to a high quality reference spectrum. When this spectrum was converted to the Kubelka-Munk (K-M) format, a remarkable increase in the *apparent* SNR is observed, see Figure 6A. The noise suppression is due to the $(1-R)^2$ form of the numerator of the K-M function and the fact that the denominator is essentially constant for weak bands. A comparison of this spectrum and a reference spectrum of phenanthrenequinone prepared both as a more intense DR spectrum (Figure 6B) and as a KBr disk (Figure 6C) indicates the success of this measurement.

ACKNOWLEDGEMENTS

This work was partially supported by the University of California Toxic Substances Research Program and through Cooperative Agreement CR812258-02 between the University of California, Riverside and the U.S. Environmental Protection Agency, Environmental Monitoring Systems Laboratory, Las Vegas. The generous donation of the Model 1800 FT-IR spectrophotometer by the Perkin-Elmer Corporation and the IR-plan microscope by Spectra-Tech, Inc. is also gratefully acknowledged.

REFERENCES

1. K. H. Shafer, P. R. Griffiths, and R. Fuoco, *J. High Res. Chromatogr. Chromatogr. Comm.*, 9: 124-126 (1986).
2. K. H. Shafer, P. R. Griffiths, and R. Fuoco, *Anal. Chem.*, 58: 3249-3254 (1986).
3. K. H. Shafer, S. L. Pentoney, and P. R. Griffiths, *Anal. Chem.*, 58: 58-64 (1986).
4. S. L. Pentoney, K. H. Shafer, P. R. Griffiths, and R. Fuoco, *J. High Res. Chromatogr. Chromatogr. Comm.*, 9: 168-171 (1986).
5. S. L. Pentoney, K. H. Shafer, and P. R. Griffiths, *J. Chromatogr. Sci.*, 24: 230-235 (1986).
6. P. R. Griffiths, and D. E. Henry, *Progress in Analytical Spectroscopy*, 9: 455-482 (1986).
7. K. Jinno, and C. Fujimoto, *J. High Res. Chromatogr. Chromatogr. Comm.*, 4: 532 (1981).
8. K. Jinno, C. Fujimoto, and Y. Hirata, *Appl. Spectrosc.*, 36: 67 (1982).
9. K. Jinno, C. Fujimoto, and D. Ishii, *J. Chromatogr.*, 239: 625.
10. K. Jinno, *Spectrosc. Letters, 1982*, 14: 659 (1982).
11. D. Kuehl, and P. R. Griffiths, *J. Chromatogr. Sci.*, 17: 471-476 (1979).
12. D. Kuehl, and P. R. Griffiths, *Anal. Chem.*, 52: 1394-1399 (1980).
13. C. M. Conroy, P. R. Griffiths, P. J. Duff, and L. V. Azarraga, *Anal. Chem.*, 56: 2636-2642 (1984).

14. C. M. Conroy, P. R. Griffiths, and K. Jinno, *Anal. Chem.*, 57: 822-825 (1985).
15. P. R. Griffiths, and C. M. Conroy, *Adv. Chromatogr.*, 25: 105-138 (1986).
16. K. Kalasinsky, J. A. S. Smith, and V. F. Kalasinsky, *Anal. Chem.*, 57: 1969-1974 (1985).
17. J. J. Gagel, and K. Biemann, *Anal. Chem.*, 11: 2184 (1986).
18. W. H. McFadden, D. C. Bradford, D. E. Games, and J. L. Gower, *Amer. Lab.*, 9: 5 (1977).
19. W. H. McFadden, H. L. Schwartz, and S. Evans, *J. Chromatogr.*, 122: 389 (1976).
20. W. H. McFadden, *J. Chromatogr. Sci.*, 17: 2 (1979).

15

Industrial Problem Solving Using Microvibrational Spectroscopy

NORMAN R. SMYRL, RANDALL L. HOWELL, DOYLE M. HEMBREE, JR., AND JOSEPH C. OSWALD
Martin Marietta Energy Systems, Inc., U.S. Department of Energy, Oak Ridge, Tennessee

15.1 INTRODUCTION

The purpose of this chapter is to describe how we are utilizing microvibrational spectroscopy (both infrared and Raman) as industrial analytical problem-solving tools at our facility. First, for perspective, we present a brief overview of the purpose and mission of the Oak Ridge Y-12 Plant and our function as a laboratory development department in support of this mission. The Y-12 Plant is a nuclear weapons component manufacturing facility operated for the U.S. Department of Energy by Martin Marietta Energy Systems, Inc. As such, the plant is involved in a wide variety of chemical and metallurgical processes in support of the manufacturing of these components. The Laboratory Development Department at Y-12 has a twofold mission:

•Provide analytical support to the Y-12 Plant in the areas of

1. development of new materials and processes,
2. troubleshooting existing processes,
3. quality assurance of materials utilized by the plant,
4. quality evaluations involved in weapons teardowns,
5. methods development to support manufacturing activities and regulatory compliance, and
6. technical expertise to support other Plant Laboratory activities.

•Provide limited analytical support to other Energy Systems facilities (Oak Ridge National Laboratory, Oak Ridge Gaseous Diffusion Plant, and Paducah Gaseous Diffusion Plant).

Our particular group within the Laboratory Development Department is primarily a molecular spectroscopy group with nuclear magnetic resonance (NMR), Fourier transform infrared (FTIR), and Raman instrumentation. Our work is focused principally in the first four areas listed above. In the production of weapons and weapons components, great emphasis is placed on quality assurance in maximizing both the dependability and lifetime of the weapons. It will become apparent as various applications are discussed in this paper that the majority of our work can eventually be related to this single issue of quality assurance. Both our FT-IR and Raman instruments possess microanalysis accessories.

Our FT-IR instrument is a Digilab (Cambridge, MA) Model FTS-15 consisting of two optical benches, one of which is permanently fitted with a universal microtransmission/reflection accessory. Our Raman instrumentation consists of an Instruments SA (Metuchen, NJ) Model U-lOOO fitted with a Nachet microscope.

15.2 MICRO-FOURIER TRANSFORM INFRARED SPECTROSCOPY

Rather surprisingly, one of our major uses of the FT-IR microscope accessory has been the qualitative macroanalysis of various types of cured polymers and plastics. Even though we are using it in the macro-analysis sense, the success of the technique still relies on ability and ease in handling small samples. It is relatively easy using a razor blade or knife to obtain a minute specimen from a polymer thin enough to analyze by transmission on the microsampling accessory. In the few cases in which the sample is still too thick, it can be placed in a KBr pellet die or a diamond anvil and pressed to a thickness suitable for analysis.

An example of this type of macropolymer analysis is shown in Figure 1. The lower curve is the spectrum for a clear, flexible tubing used on the air bearings of a dimensional inspection machine. This particular sample, which was characterized as plasticized PVC, had ruptured under the 80-psi pressure used to supply air to the bearings. On inspection of this machine, other weak points with further potential to rupture were noted for the tubing. Similar tubing on an adjacent machine, which apparently was rated properly for the pressure conditions, was sampled and identified as a different type of tubing composed of polyurethane (the upper spectrum in Figure 1). An important point is that the size of the sample required for this type of characterization did not in any way compromise the operation of the machine or the structural integrity of the tubing. For this particular problem, a recommendation was made to replace all the suspect tubing with a polyurethane type that was properly rated for the pressure conditions.

Another example of macropolymer analysis is shown in Figure 2 for the characterization of certain materials from a filter used in a recrystallization process. It is important from the standpoint of the quality and purity of the recrystallized material that all components related to the process be characterized for their overall compatibility. The upper curve of Figure 2 is the spectrum of the material that comprises a support ring for the filter, which was identified

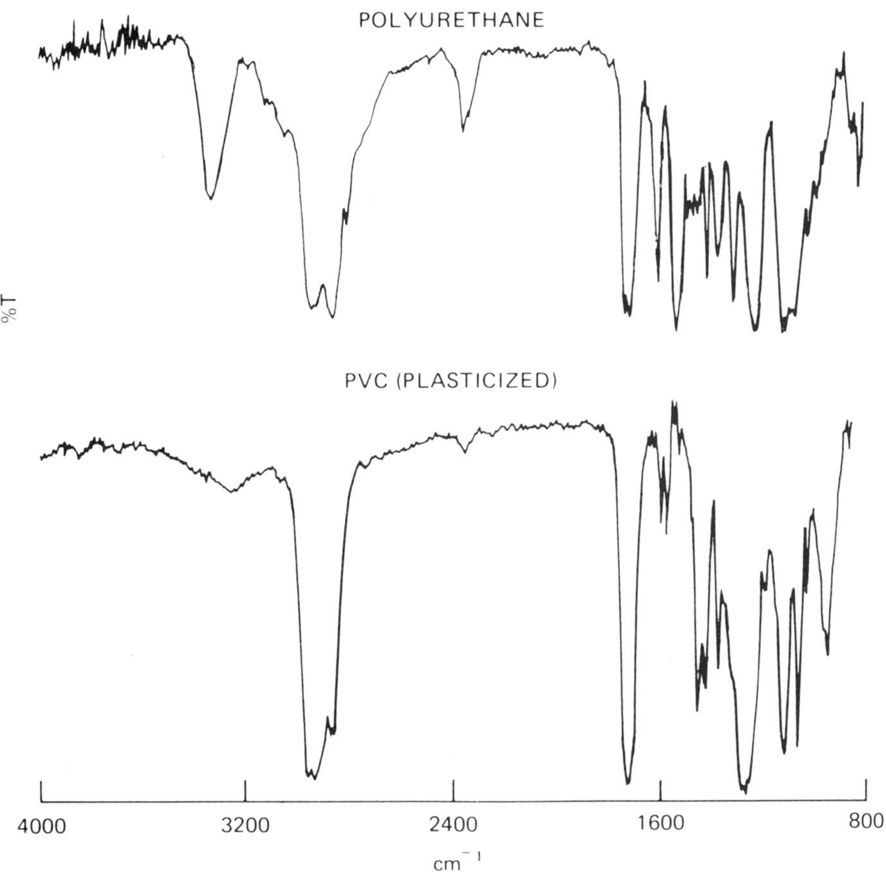

Figure 1. Infrared macropolymer analysis: infrared spectra of flexible tubing from a dimensional inspection machine.

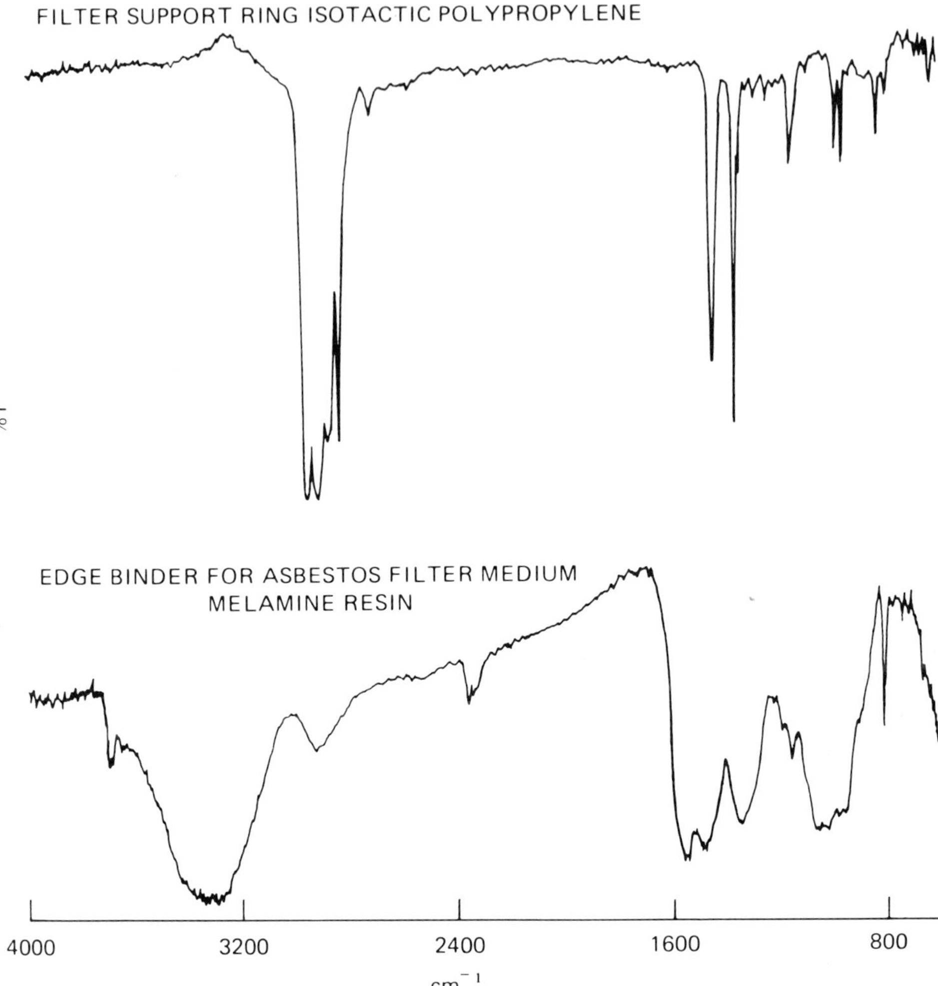

Figure 2. Infrared macro polymer analysis: infrared spectra of materials from filter used in a recrystallization process.

as isotactic polypropylene. The lower curve is the spectrum of the edge binder for the filter, which was determined to be a melamine resin. The spectrum is also observed to contain features from the asbestos filter medium, as indicated by the bands at 3700 and 1000 cm^{-1}.

We also use the infrared microscope accessory in the more conventional sense in our problem-solving efforts by examining microscale samples such as fibers, particulates, and inclusions. An example of this type of microanalysis is shown in Fig. 3. Here the technique has been used to characterize the fiber support for an abrasive-type wool used in a cleaning operation involving a particular type of weld. Again, because it is important from a quality standpoint that the cleaning process itself does not leave contaminants on the surface, it is necessary to characterize all materials used in the process. The upper curve is the spectrum of a single fiber that was flattened to a suitable thickness in a KBr die. The lower spectrum is that for a standard nylon 6/6 sample obtained on the microscope stage. The two spectra are essentially identical.

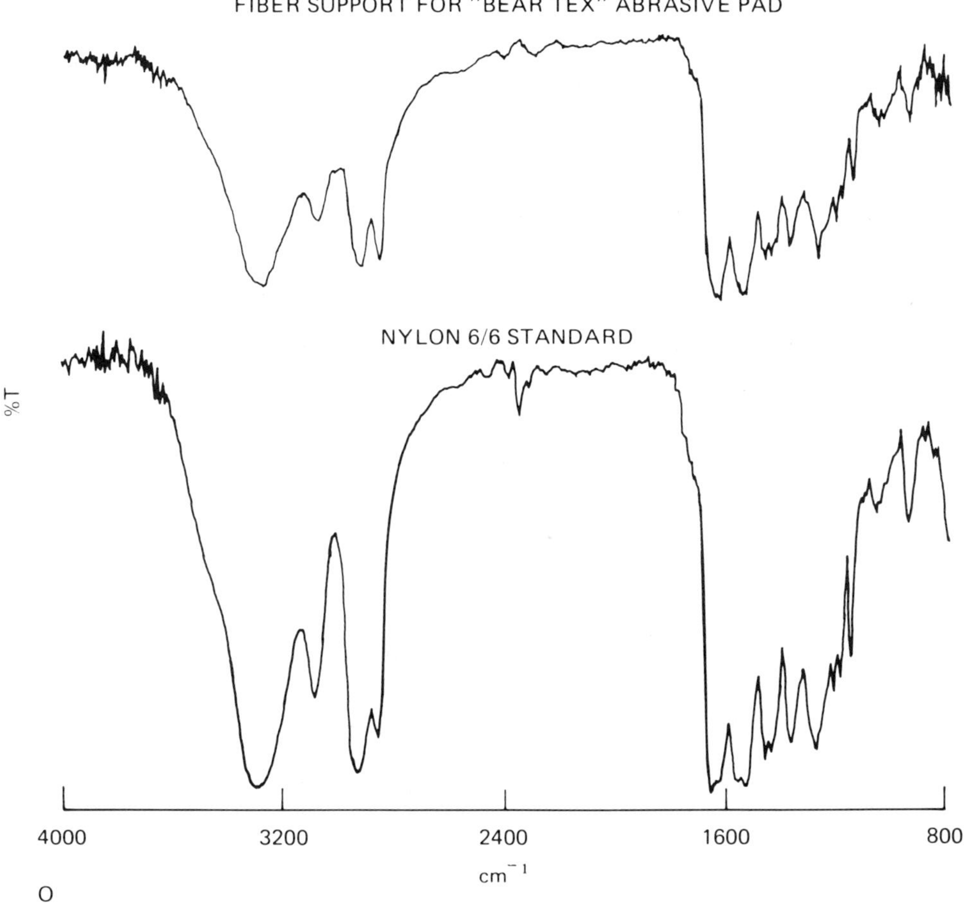

Figure 3. Infrared micro polymer analysis: infrared spectra fiber support for "Bear Tex" abrasive pad (upper) and nylon 6/6 standard (lower).

Table 1 lists various types of plastics identified using the FT-IR microscope from the filter debris of a water-soluble material used in fabricating a certain part. These particles ranged from ~0.5 mm down to 25 μm in diameter. Most of the materials identified are traceable to plastic bearings, bushings, seals, and the like for various components in the processing chain for this particular material. This information is useful in evaluating potential and unusual outgassing characteristics of the material. It should be pointed out in relation to the analysis described above that the materials that comprise the assembly, and the assembly itself, must meet stringent outgassing limits before the components or the assembly can be certified.

Plastics and polymers that contain opaque filters, such as carbon black, are difficult to characterize by infrared. The carbon filler, particularly at moderate to high loading, cause great difficulty for most IR sampling methods. Some success has been obtained in studying these materials by transmission IR using microtoming techniques. Another approach that can be used is pyrolysis providing the concentrations of plasticizers and other volatile components are low. Figure 4 shows an example of the characterization of a black foam particulate (from the same filter debris shown in Table 1) using a micro-pyrolysis technique. The technique consists of transferring the particle to a small, very shallow platinum boat and placing a KBr pellet or window of an appropriate size over the boat. The particulate is then pyrolyzed in a conventional pyrolysis accessory. With the aid of the microscope, it is usually

Table 1. Infrared micro polymer analysis of particle debris collected from filtration of H_2O; soluble material used in fabricating weapons parts.

PARTICLE IDENTIFICATIONS	
Teflon polyisoprene	
PVC (plasticized)	Silicon rubber
Nylon fibers	Polyester fibers
Cellulose fibers	Poly (methyl methacrylate)
Epoxy	Cellulose acetate butyrate
Maleic resin (UNI-REZ)	Styrene/butadiene copolymer
Acrylate/styrene/acrylic	Polyurethane
Acid terpolymer	Melamine (plasticized)
Polystyrene	Polyvinyl alcohol
Polyethylene	

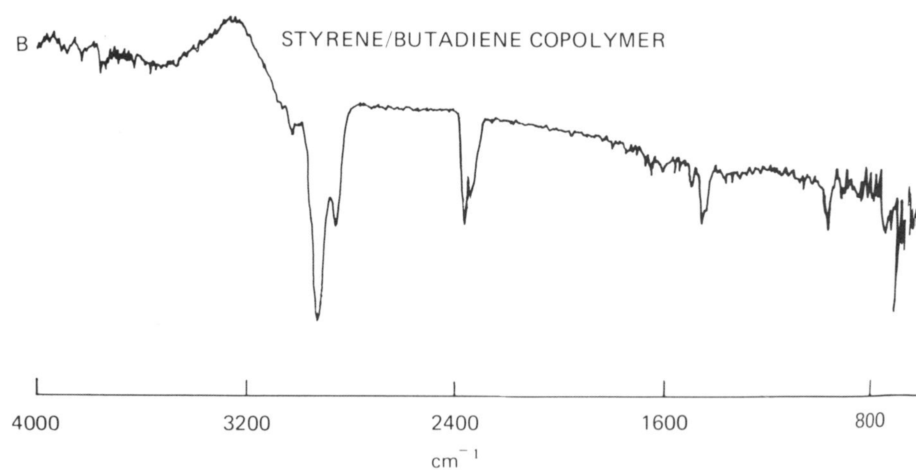

Figure 4. Infrared spectra of pyrolyzate of black foam particle from filter debris.

fairly easy to locate the pyrolysis condensate. The particulate depicted in Figure 4 was pyrolyzed in this manner, and the resulting condensate appears to be composed of a styrene/butadiene copolymer.

As mentioned previously, we also have reflection measurement capabilities with our FT-IR microscope accessory and have used it to a limited extent in our problem-solving activities. Figure 5 shows a study using this capability with regard to oil coverage on uranium machine chips or turnings. A recent change in machine coolants from a 60-40 mixture of perchlorethylene/mineral oil to an aqueous-based coolant consisting of a 50-50 mixture of ethylene glycol and water and 0.5% borax for machining enriched uranium initially resulted in an increased incidence of chip fires. The oil present in the previous coolant tended to form a protective coating on the chips, thus helping prevent oxidation and fire risks associated with the machining process. The principal step at which there appears to be the greatest potential for fires with the new coolant is in the Freon cleaning step to remove water-based coolant from the wet chips. Some experiments were carried out to determine if oil coverage of the chips could be accomplished during Freon evaporation in this chip-cleaning step to subsequently help reduce fire risk in recycling the chips. The FT-IR microscope was used to profile the width of both sides of turnings ~1 mm in width and 25 μm thick to determine the extent of oil coverage following Freon evaporation. The C-H stretching region (2700-3100 cm^{-1}) was monitored as shown in Figure 5 for solutions of Freon containing 2300 and 500 ppm of oil. The 2300-ppm case indicated good coverage on both sides, whereas

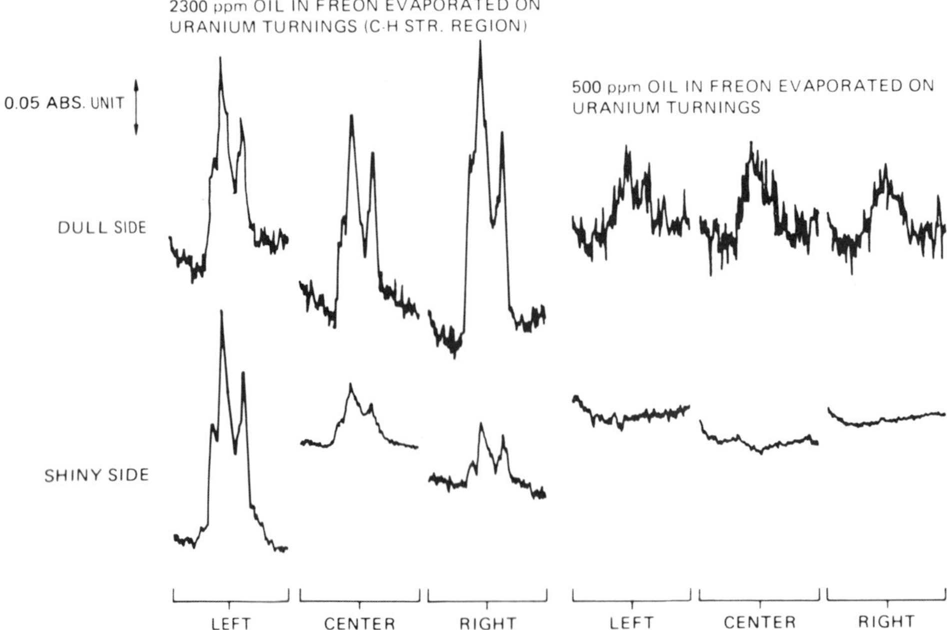

Figure 5. Infrared microreflectance: profile of oil coverage on uranium machine turnings (2700-3100 cm^{-1}).

in the 500-ppm case the oil was likely present but undetectable on the shiny side using the near-perpendicular incidence geometry of the microscope. Two general observations can be made from these data: (1) because the dull side is the outside portion of the machine turning, which is considerably rougher, it takes up more oil than the smoother shiny side, and (2) evaporation of the Freon tends to occur from the middle of the chips to the edges, leaving a higher concentration of oil near the edges.

Another example of our utilization of the microreflectivity capability is shown in Figure 6. A 1/8 in. (3.175 mm) metal bellows suspected of having a surface contaminant was analyzed both on the inside and outside folds of the bellows. The closest match of the contaminant spectrum to those available in our standard spectral files was for a Kel-F poly (chlorotrifluoroethylene) type of grease. [However, it was pointed out to us at the FACSS meeting in St. Louis that a Krytox (polyhexafluoropropylene epoxide) grease is more likely the proper identification.]

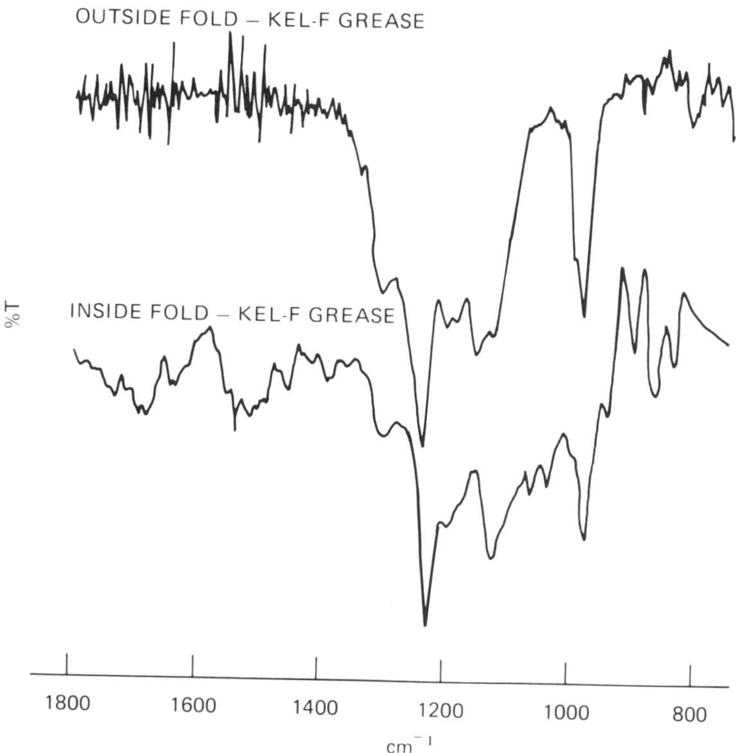

Figure 6. Infrared microreflectance: infrared spectra of contaminant on folds of 1/8" metal bellows.

15.3 MICRO-RAMAN SPECTROSCOPY

Although we have only had our Raman instrument in operation for a little over 8 months, we have begun to integrate this technique into our problem solving efforts and are finding many situations well-suited to the microanalysis capabilities of this technique. Figure 7 shows one example in which a contaminant on a uranium machine turning was characterized using this technique. The lower curve is the Raman spectrum of the surface contaminant, which was identified as boric acid. This contaminant likely results from conversion of the borax (sodium tetraborate) to boric acid in the water/ethylene glycol machine coolant (discussed previously in the infrared section) that is utilized in machining enriched uranium. Boron is a neutron absorber and is added to the coolant in the form of borax to reduce the risk of a criticality accident. Attempts are made to remove all boron that accumulates on the chips before they are recycled; consequently, it is important to know as much as possible about the chemistry of the boron, both in solution and on the chip surface.

Figure 8 illustrates an application of Raman microanalysis to the characterization of a pyrocarbon-coated experimental fuel pellet that contains ThO_2 as the fuel. The photomicrograph cross section of the fuel pellet shows five distinct regions. The first three represent different areas within the carbon coating (~250 µm in thickness) as indicated by the corresponding Raman spectra for these regions. Bands are observed at 1574 and 1350 cm^{-1}, which have been attributed in prior literature studies to crystalline graphitic-type carbon and an amorphous-phase carbon, respectively [1 - 3]. It appears that the amount of the amorphous-phase in the coating varies considerably throughout but has the highest concentration in the interior region of the coating. The inner two regions of the photomicrograph are due to the sol-gel ThO_2 bead having a diameter of ~250 µm. Both of these regions yield spectra with a characteristic band for the cubic form of ThO_2 at 466 cm^{-1}.

One of our customers from Oak Ridge National Laboratory submitted a clear crystal of SiC for characterization of black inclusions within the crystal. Figure 9 shows the Raman spectra of both the SiC and the inclusions. The SiC is in the ß (cubic) form with characteristic bands at 765, 786, and 963 cm^{-1} [2]. The inclusion can be identified as graphitic carbon, which is totally in the crystalline phase, as indicated by the observation of only the 1574 cm^{-1} band. One can also see some residual bands due to the SiC and two broad bands between 1000 and 1200 cm^{-1}, which are due to instrumental artifacts.

Figure 10 shows the Raman spectrum of the abrasive material from the wool used in a weld-cleaning operation, which was mentioned previously in the infrared discussion. The abrasive material was identified as SiC in the α (hexagonal) form [3]. It should be noted that the two forms of SiC examined in this figure and in Figure 9 have well-characterized phases with distinctly different Raman spectra.

One final area of discussion concerning micro-Raman analysis involves electrochemical plating processes, of which there are various types in use in the Y-12 Plant. The chemistry associated with these processes is often more of an art than a science. A Raman electrochemical cell for use in conjunction with the microscope stage has been designed and constructed to perform some basic studies involving these plating processes [5]. It is felt that a better understanding of the chemistry associated with each process will allow better control of the process and will permit a more scientific analysis of problems to be made when they occur. Figure 11 is a diagram of this cell. The solution cavity, which contains the working electrode, the Ag/AgCl reference electrode, and the

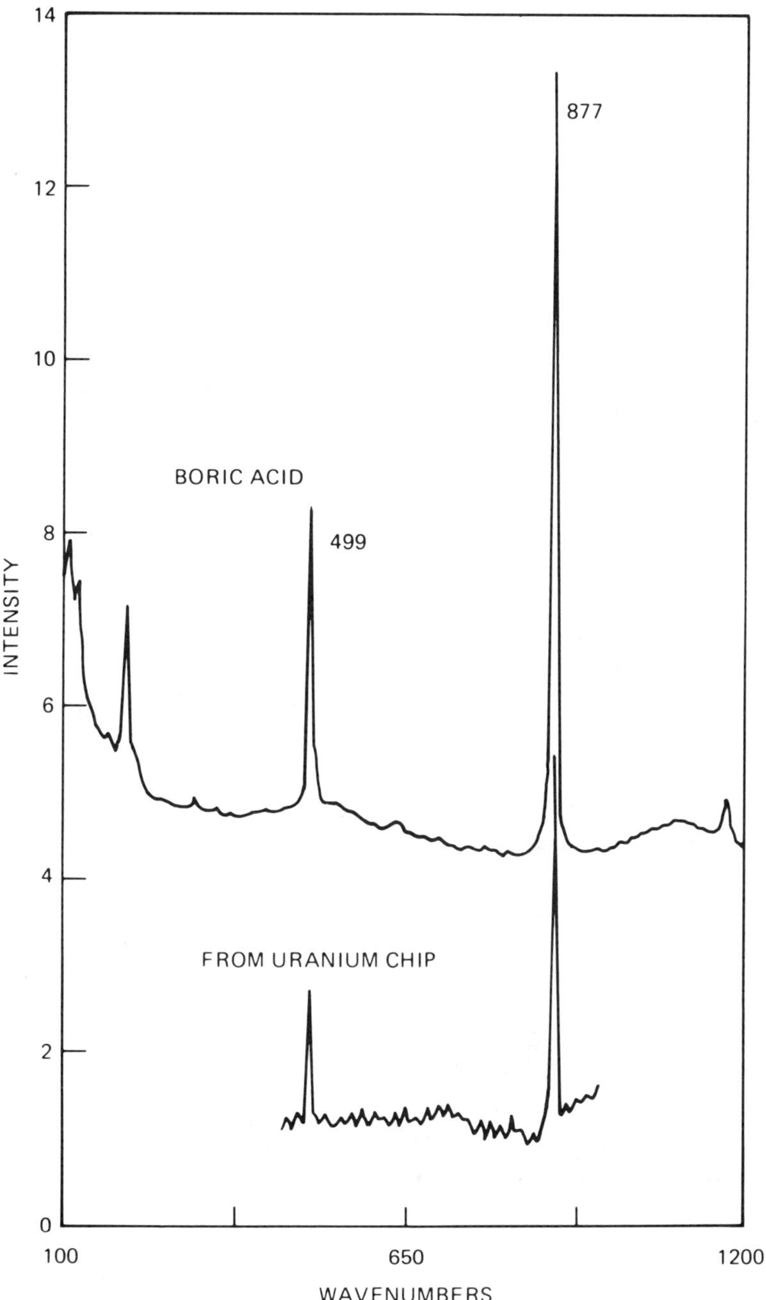

Figure 7. Raman spectra (lower) contaminant particle on the surface of uranium machine turning (upper) boric acid.

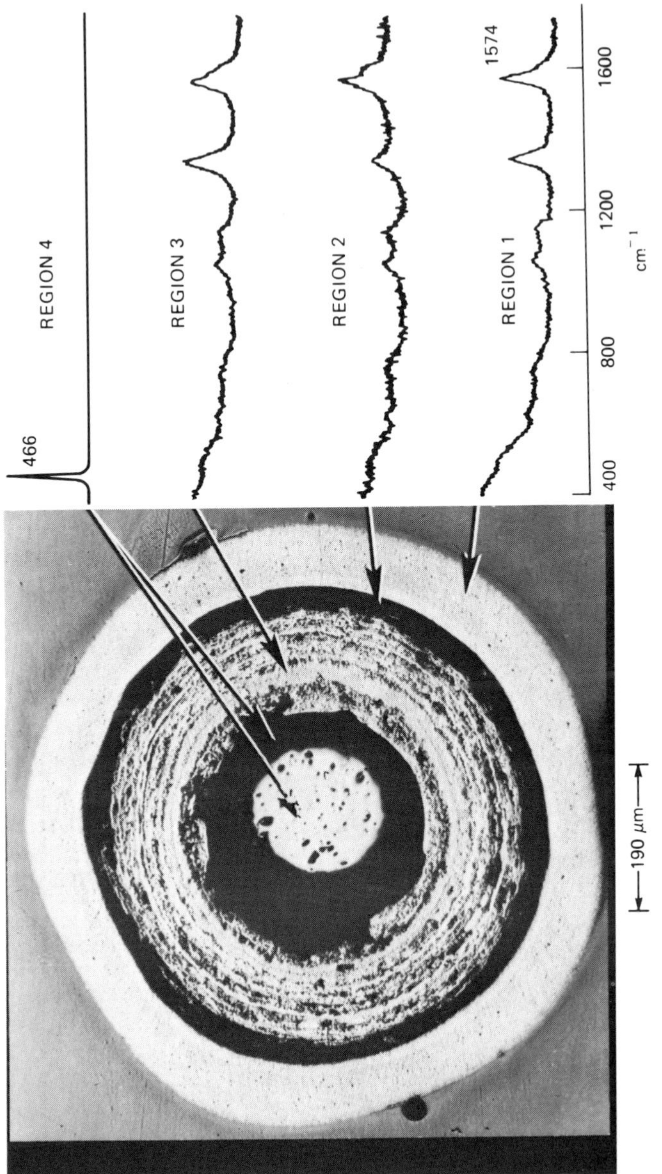

Figure 8. Raman microanalysis: Raman spectra of different regions of thoria nuclear fuel pellet cross section.

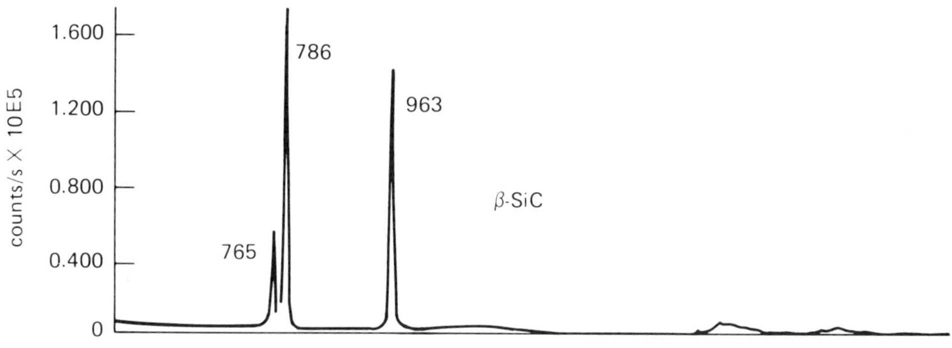

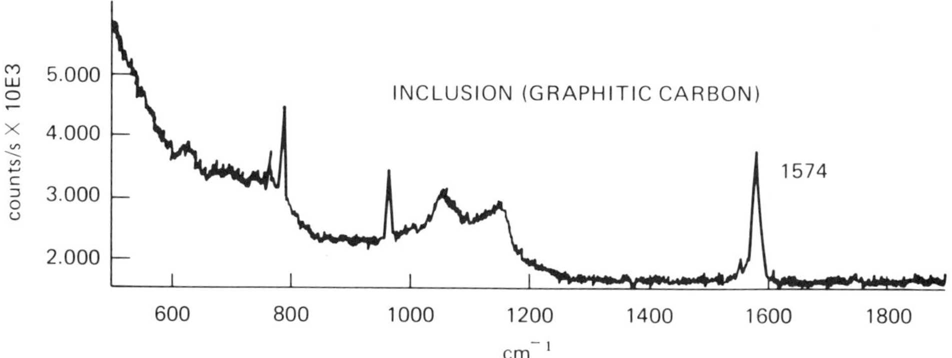

Figure 9. Raman microanalysis: Raman spectra of SiC crystal containing carbon inclusions.

Pt counter electrode appear in the right side of this figure. The cavity is covered by a thin glass coverslip that creates a gap of ~0.5 mm between it and the working electrode.

Some of the important characteristics of this cell when used in conjunction with the Raman microscope include:

1. exchangeable working electrodes,
2. low sample volume (1-6 mL),
3. visual examination of the electrode surface using the Raman microscope,
4. spatial resolution at the laser focal point (~1 µm),

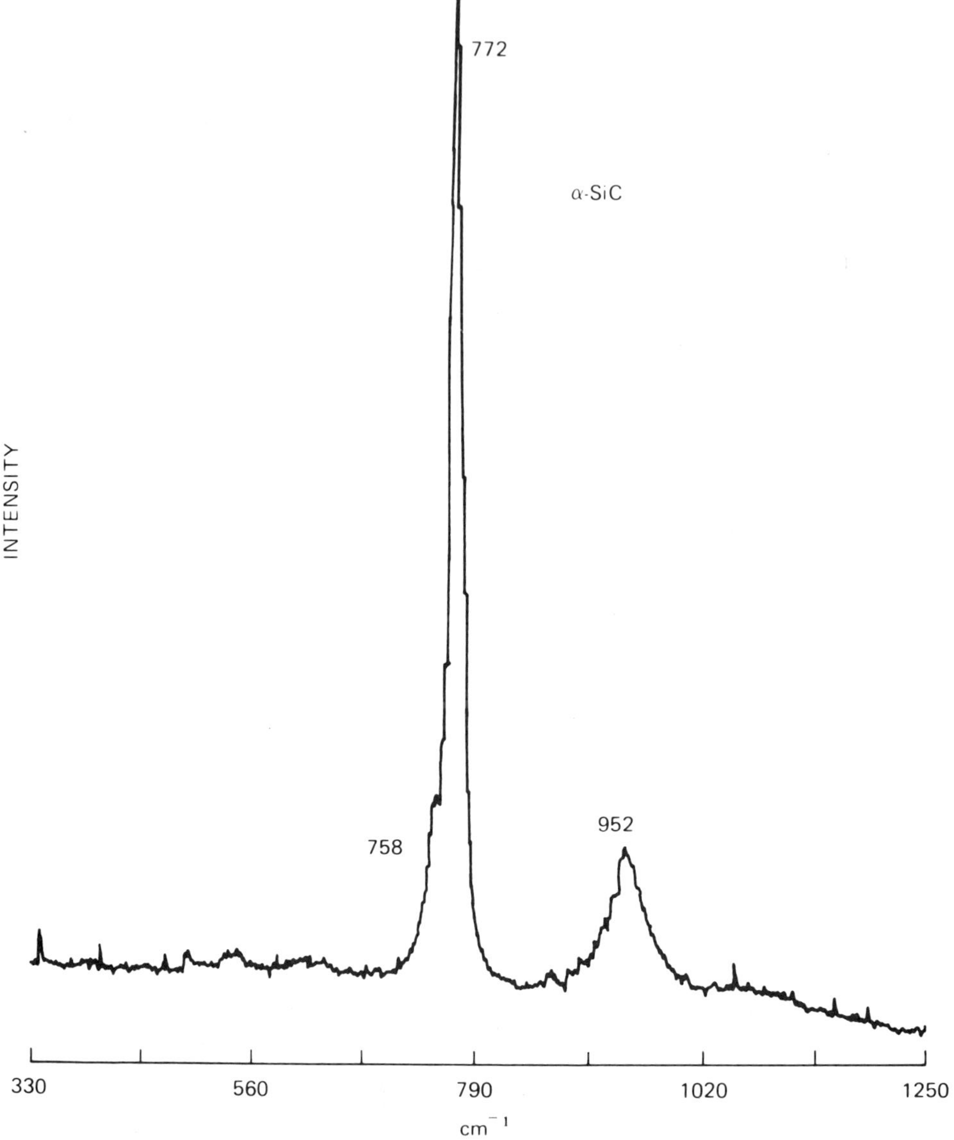

Figure 10. Raman microanalysis: Raman spectrum of abrasive particles on "Bear Tex" wool.

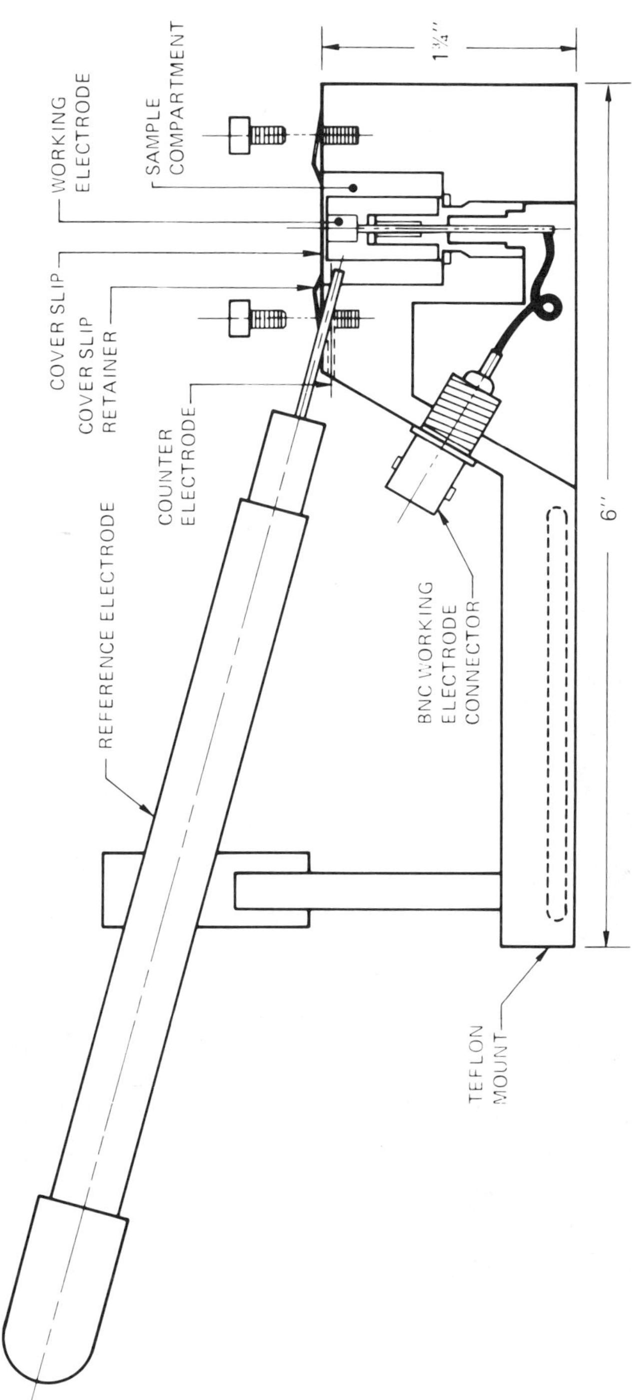

Figure 11. Electrochemical cell for Raman microspectroscopy.

5. essentially a thin-layer electrochemical cell,
6. Raman signal generation primarily from the laser focal point, and
7. accurate placement of the laser focal point made possible by calibration of the microscope fine adjustment.

Figure 12 is a photograph of the cell in use with the Raman microscope. We have used this cell to study surface-enhanced Raman effects [6, 7] of certain model systems, such as pyridine on various electrode surfaces, including Ag, Cu, and Ni. Figure 13 illustrates a study involving $KAu(CN)_2$, which is a commercially important starting material in gold electroplating. Surface-enhanced Raman spectroscopy was used to investigate various steps in the plating process. The 2166 cm^{-1} band is due to the $KAu(CN)_2$ complex in solution, and no surface-enhanced effects are observed at zero potential. However, as the potential is stepped negatively, a band is observed at 2122 cm^{-1}, which is due to a surface-associated gold cyanide complex. The band reaches a maximum in intensity at -0.32 V and begins to diminish and

Figure 12. Photograph of the electrochemical cell in use with the Raman microscope.

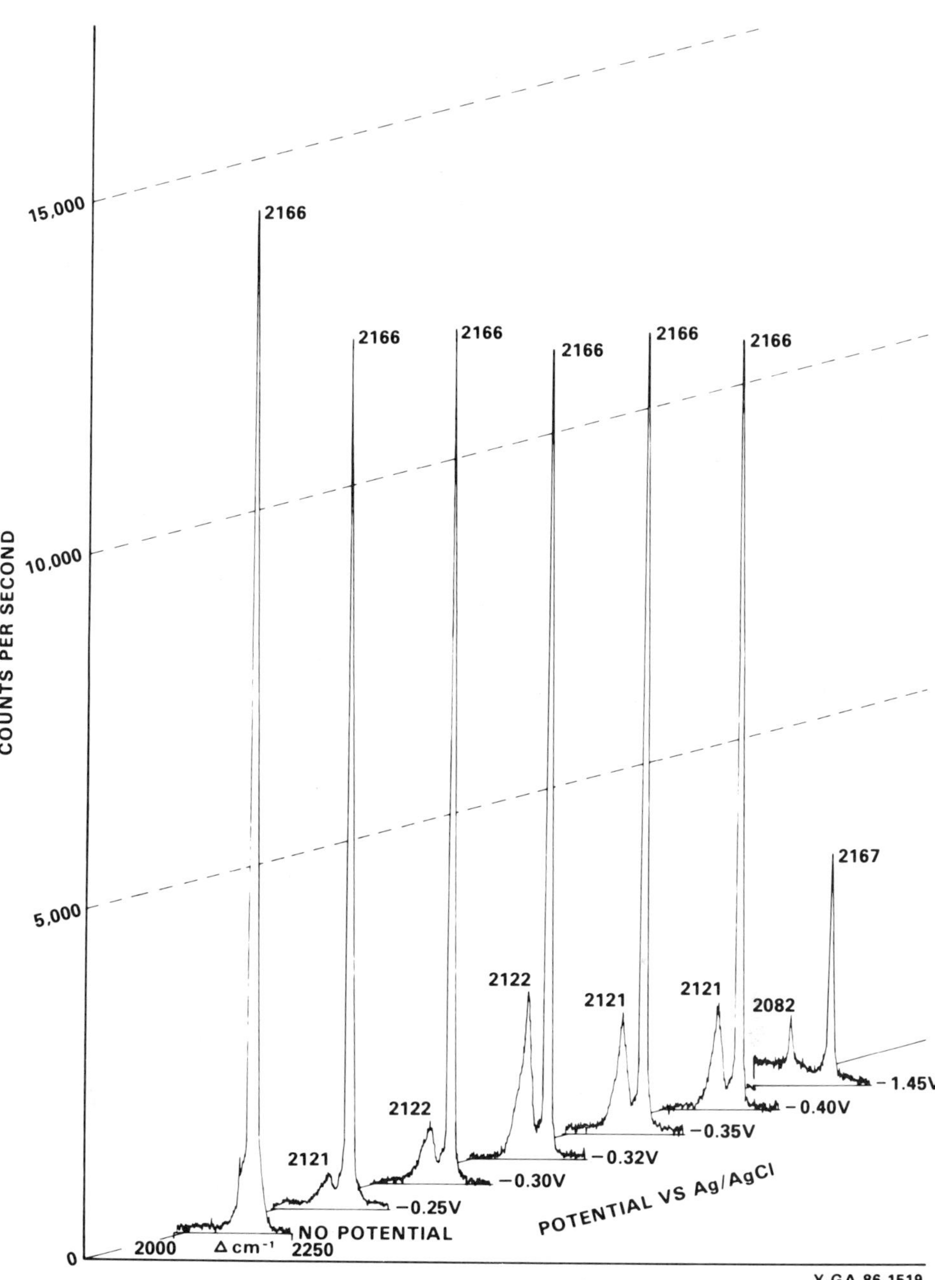

Figure 13. Electrochemical surface-enhanced Raman spectral (SERS) study of gold electroplating system on a silver electrode.

finally disappears completely at -1.45 V as gold begins to plate onto the silver electrode. Finally, a band due to free cyanide at 2082 cm^{-1} is observed. Because cyanide does not show surface-enhanced effects on gold with 514.5 nm excitation, the surface-enhanced effect is passivated as the gold is plated. This particular set of data represents only some preliminary work but demonstrates the potential of this technique related to the study of various plating chemistries.

15.4 CONCLUSIONS

It can be concluded that infrared and Raman microspectroscopic techniques represent a new set of problem-solving tools available to the industrial analytical spectroscopist. These techniques expand the scope of problems that can be attacked by vibrational spectroscopy. Infrared microspectroscopy appears to be a very strong technique in the areas of both macroanalysis and microanalysis of organic polymers. Raman microspectroscopy, on the other hand, appears to be the stronger of the two techniques in the area of inorganic analysis. Eventually, with the establishment of an extensive data base, Raman could supplant X-ray diffraction in certain areas of inorganic analysis because it is a much faster technique and also has the ability to provide characterization for amorphous phases.

REFERENCES

1. W. L. Johnson III, in *Microbeam Analysis* (A. D. Romig, Jr. and W. F. Chambers, eds.), San Francisco Press, San Francisco, CA, pp. 26-28 (1986).
2. T. C. Chieu, and M. S. Dresselhaum, *Phys. Rev. B*, 26(10): 5867 (1982).
3. F. Tuinstra, and J. K. Koenig, *J. Chem. Phys.*, 33(3): 1126 (1970).
4. P. Krautwasser, G. M. Begun, and P. Angelini, *J. Am. Ceramic Soc.*, 66(6): 424 (1983).
5. D. M. Hembree, J. C. Oswald, and N. R. Smyrl, *Appl. Spectrosc.*, 41 (2): 267 (1987).
6. M. Fleischmann, P. J. Hendra, and A. J. McQuillan, *Chem. Phys. Lett.*, 26: 163 (1974).
7. R. K. Chang and T. Furtak, *Surface Enhanced Raman Scattering*, Plenum, New York (1982).

16

Infrared Microspectroscopic Solutions to Manufacturing and Quality Control Problems

DAVID W. SCHIERING *Perkin-Elmer Corp., Norwalk, Connecticut*

16.1 INTRODUCTION

At the 1983 Pittsburgh Conference, an infrared microsampling device was introduced that was designed specifically for Fourier transform infrared (FT-IR) spectrophotometers. The design of this device allowed the observation of the sample through a reflecting objective which also served as an infrared radiation collector. Another feature was a remote image masking aperture which permitted the simplified isolation of the area of interest. Although infrared imaging in a normal sense is not accomplished with this type of accessory, the instrument came to be known as an infrared microscope. Several other commercial infrared microscopes were subsequently introduced.

However, the birth of infrared microscopy preceded 1983. It appears that in 1949 the first infrared spectra were measured through a device employing a Cassegrainian objective with the ability to view the sample [1]. The microscope had been previously used to collect absorption spectra in the UV and visible regions. The first commercial infrared microscope was the Perkin-Elmer Model 85, introduced in 1953 [2]. The design of this microscope is remarkably similar to that of today's FT-IR microscopes. Many of the same applications of interest today were performed with this device. Spectra were recorded of fibers, crystals, and biological samples, and polarized spectra were also recorded. The Model 85 was a single beam instrument; the sampling time was relatively long, and since the infrared spectrophotometers were dispersive, the resolution varied significantly across the spectral region. The Model 85 could not be considered a commercial success. Off-axis beam condensers became more popular for microsampling due to their relatively higher performance.

The influx of commercial rapid scanning FT-IR spectrophotometers with sensitive detectors has made infrared microscopy a viable technique. More interest than ever before is now focussed on infrared microsampling. Technical sessions and symposia are now devoted to infrared microspectroscopy at major spectroscopy meetings [3]. An academic laboratory has been established which is devoted to research in infrared and Raman microspectroscopy.

This chapter will consist of a somewhat general overview of some types of problems encountered in manufacturing or quality control environments and the application of infrared microspectroscopy to their solution. Infrared microspectroscopy may be utilized to solve a wide range of problems including those in the polymer, semiconductor, electronics, paper, coatings, pharmaceutical and chemical industries. The potential application of infrared microscopy in such diverse areas has contributed greatly to the technique's popularity.

16.2 EXPERIMENTAL

The spectra reported were measured using the Spectra-Scope* or IR-plan* infrared microscope accessories. The Spectra-Scope utilizes a fixed focus, off-axis ellipsoid to condense the infrared radiation at the sample. The transmitted radiation was collected with a 15X Cassegrainian objective (N. A. = 0.58), passed through a knife-edge variable aperture, and directed to an instrument mounted detector. For sample observation, the Spectra-Scope has transmitted and reflected light illumination. A monocular viewer with a lOX eyepiece was used.

The IR-plan is an infrared microsampler designed around a research grade optical microscope. It features an on-axis infrared sampling arrangement employing a Cassegrain as a condenser and another as the infrared objective. This on-axis design allows correction for sample or substrate induced lensing effects and the implementation of dual remote image aperturing. Aperturing prior and subsequent to the sample has been shown to greatly decrease diffraction induced stray light [4]. The infrared radiation was focussed at the sample by a 15X Cassegrainian objective (N.A. = 0.58); the transmitted radiation was collected by a lOX Cassegrainian objective (N. A. = 0.71).

*Spectra-Scope and IR-plan are trademarks of Spectra-Tech, Stamford, CT.

Infrared reflectance spectra of small areas were also obtained with the IR-plan. In this arrangement, the 15X objective both focussed the incident radiation and collected the reflected radiation. The sample was placed on a 2 by 3 inch rotatable stage and observed through a binocular viewer or on a video screen. Attached to the rotating nosepiece were 4X and lOX glass objectives as well as the 15X Schwarzschild-type Cassegrainian. An Olympus PM-lOAD automatic exposure 35mm camera unit was used to record the photo-micrographs. The 35mm camera body was placed subsequent to a 2.5X photo eyepiece at the top of the trinocular head of the IR-plan. Transmitted and reflected light illumination were used. The miniature diamond anvil optical cell was from High Pressure Diamond Optics, Tuscon, AZ.

A Perkin-Elmer model 1750 or 1800 FT-IR was used in conjunction with the microscope accessories to measure the spectra. For the model 1750, a DTGS (2mm X 2mm target) detector was used with the Spectra-Scope. A microscope mounted narrow band MCT detector (.25mm X .25mm target) was used with the IR-plan. For the model 1800, an instrument mounted narrow-band MCT detector with an immersion lens and .2mm X .2mm target was utilized with the IR-plan. A narrow-band MCT detector with a lmm X lmm target was used with the Spectra-Scope. MCT detector and/or preamp induced nonlinearity [5] was corrected using a potentiometer on the preamp. The source of the model 1800 was a temperature stabilized hot wire whereas the source for the 1750 was a ceramic. Medium Beer-Norton apodization was utilized for the calculation of the spectra.

16.3 RESULTS AND DISCUSSION

A large area of application of infrared microscopy is in the analysis of polymeric materials. In many cases the analyst is interested in identifying a small area in a larger matrix. Individual layers in multilayered structures have been identified as well as imperfections which are sometimes referred to as "fisheyes" [6]. The simplicity in sample location and isolation afforded by infrared microscopy simplifies the analysis of these types of samples in comparison with infrared microsampling with beam condensers. In this discussion, the analysis of polymer imperfections will be presented.

The polymer whose structure is shown in Figure 1 was manufactured utilizing a novel process. Under microscopic examination, many small imperfections were evident (Figure 2). These imperfections were on the surface or incorporated in the bulk material. The imperfections ranged in size from 10 to 50 micrometers. In Figure 3 are shown infrared spectra of an imperfection of 30 micrometer dimensions (bottom) and of a portion of the film having no imperfection (top). The same aperture dimensions were used for the collection

$-CH_2-CH_2-C_6H_3(Cl)-CH_2-CH_2-C_6H_3(Cl)-$

Figure 1. Structure of polymer. A film of this material had many imperfections.

of both spectra. The spectra were recorded with the IR-plan and model 1800 utilizing the indicated parameters.

The spectrum of the matrix has more prominent interference fringes in the 4000-2000 cm^{-1} region and a relative intensity difference (compared to the imperfection spectrum) in the 1200-1100 cm^{-1} region. Otherwise, the spectra correspond. It appears that the data is inconsistent with a chemical disparity

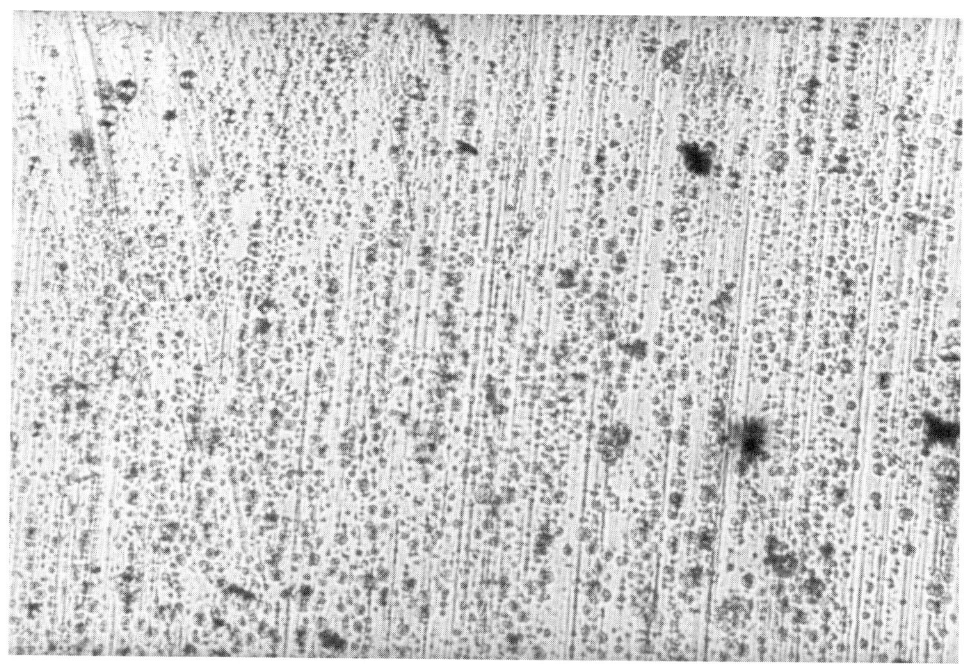

Figure 2. Photomicrograph of imperfections in a polymer film whose structure is shown in Figure 1.

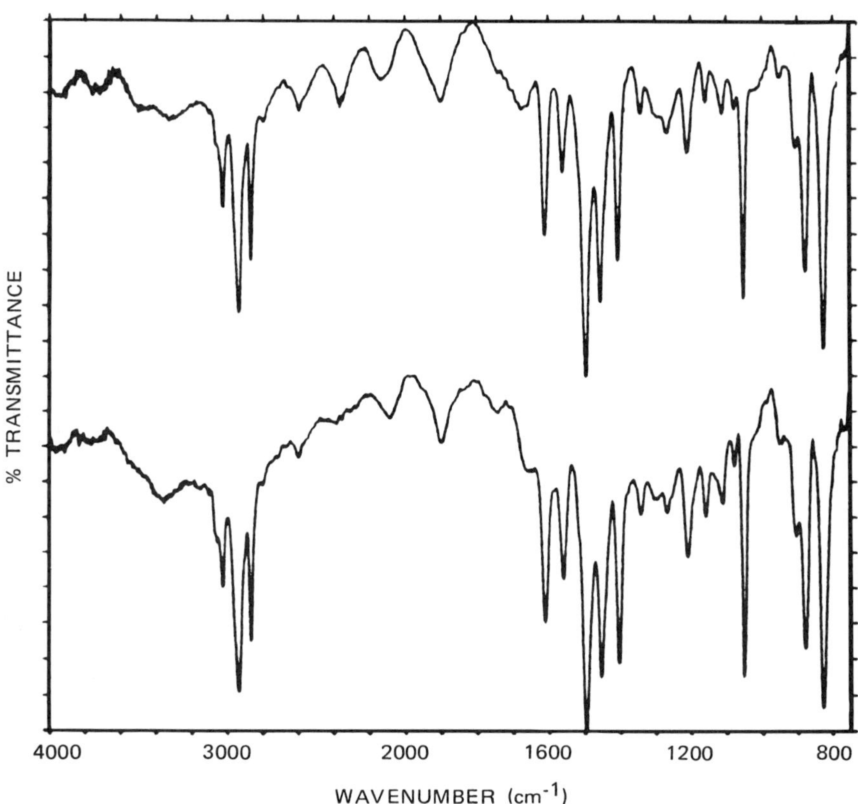

Figure 3. Top: polymer matrix. Bottom: imperfection area. Conditions: model 1800, IR-plan microscope, narrow-band MCT detector (immersed, .20mm X .20mm element), 8 cm-1 resolution, 128 scans.

between the imperfection and matrix areas. Both spectra show evidence of water. The frequency of these water bands are consistent with liquid water; therefore, detector icing (crystalline water has a band near 3240 cm^{-1}) was discounted as a source. The possibility of absorption bands outside of the indicated frequency range was discounted from knowledge of the manufacturing process. After sampling a statistically significant number of imperfections and observing the same result, it was concluded that all of the defect areas were of the same chemical identity as the matrix.

The spectra in Figure 4 are also of an imperfection and a polymer matrix. The matrix is polyethylene. The 250 micrometer imperfection is clearly of a

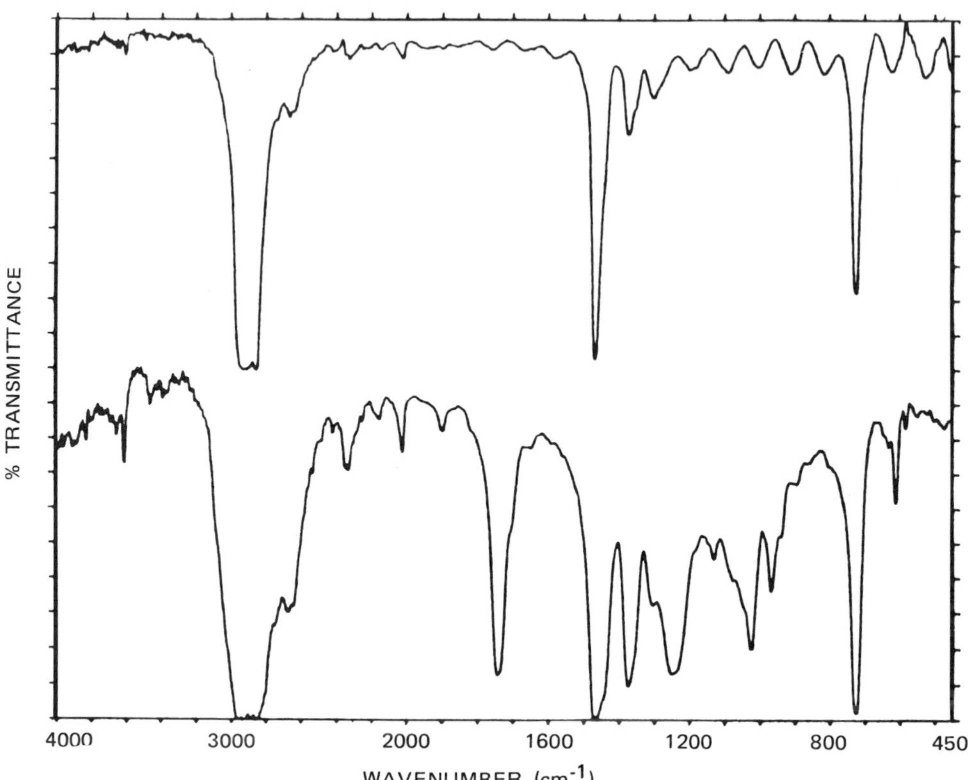

Figure 4. Top: polyethylene matrix, Bottom: imperfection. Conditions: model 1750, 4cm-1 resolution, DTGS detector, 32 scans, Spectra-Scope microscope.

different chemical makeup than the matrix and was identified as an ethylene/ vinyl acetate copolymer. As is apparent from the very acceptable signal-to-noise ratios obtained with relatively few scans, a less sensitive DTGS detector is all that is required for sample sizes of this order.

Silicon wafers require an extremely clean environment for their production and processing. However, small amounts of organic impurities may become imbedded in or reside on the surface of the wafers. Possible sources of organic impurities include photoresist processing chemicals, chemicals present in the water supply, and particles from handling. These impurities are undesirable and reduce the production yield of integrated circuit devices. Infrared microspectroscopy may be utilized to elucidate the source and correct

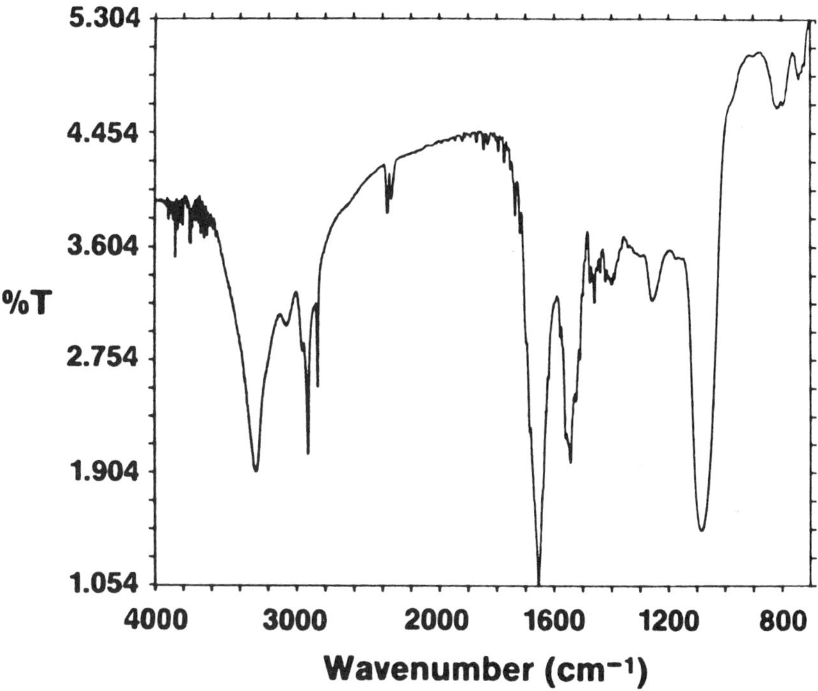

Figure 5. Spectrum of a contaminant particle recorded through a silicon wafer. Conditions: model 1800, Spectra-Scope microscope, narrow-band MCT detector (1mm X 1mm element), 4 cm-1 resolution, 64 scans.

the problem. In this example, a 78 X 268 micrometer particle was located by the optical microscope on the surface of a silicon wafer. Reflected light illumination was necessary as the wafer did not transmit visible light. A remote image of this particle was masked and its infrared spectrum recorded by transmittance through the wafer (Figure 5). An instrument mounted narrow-band MCT detector with a 1 X 1 mm target was used in this example. The strong Si-O stretching absorption band intrinsic to the wafer is evident in the spectrum near 1100 cm^{-1}. The particle was then translated out of the field of view and a spectrum of the silicon wafer was collected using the same aperture dimensions used for the particle. The silicon wafer absorptions are compensated in the difference spectrum corresponding to the particle (figure 6). The bands near 3280, 1660, and 1550 cm^{-1} are characteristic of a secondary amide and the impurity appears to be a protein particle, presumably a skin flake accidentally deposited during handling.

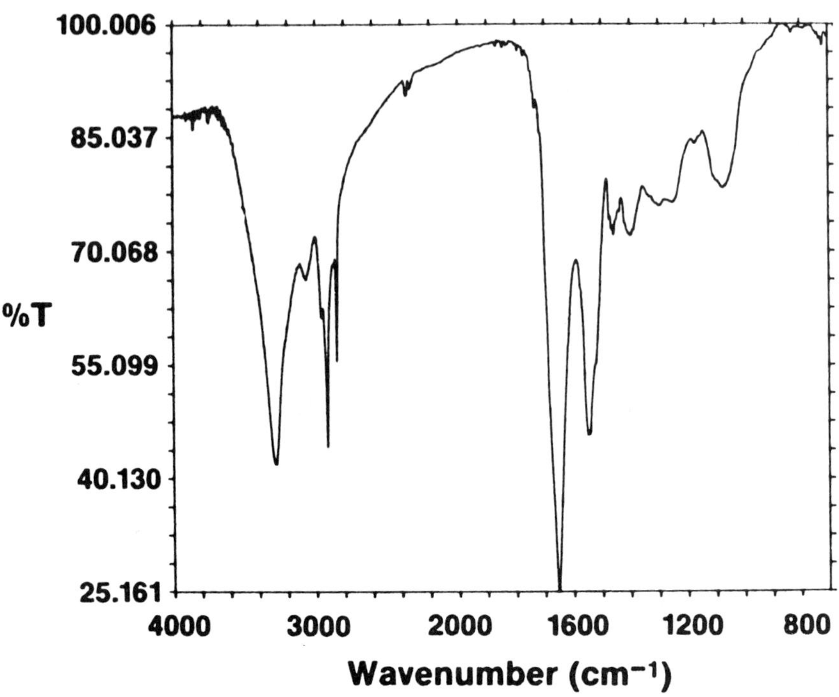

Figure 6. Difference spectrum corresponding to the contaminant on the surface of the silicon wafer.

Most infrared microscopes allow the capability of recording reflectance spectra. This is a definite advantage over beam condensers which commonly do not have this feature. For many samples, it is not practical to remove the suspect area from the matrix, as discussed. For those on metal substrates or other surfaces exhibiting reflectivity, reflection-absorption spectra can be measured. Figure 7 is a photomicrograph of an 40 X 66 μm imperfection area on a Winchester-type magnetic disk. This area was the result of an anomaly in the manufacturing process. The active film of the disk is produced by spin coating an iron (III) oxide resin dispersion over an aluminum substrate. The resin is then cured by heating.

Figure 8 is a comparison of the reflectance spectra obtained from the imperfection area and an adjacent area of the disk (the same aperture was utilized in both cases). The spectrum of the bulk film of the magnetic disk is characteristic of an aromatic polyurethane. Obvious inconsistencies exist between the spectrum of the imperfection area and the bulk coating. The 2335

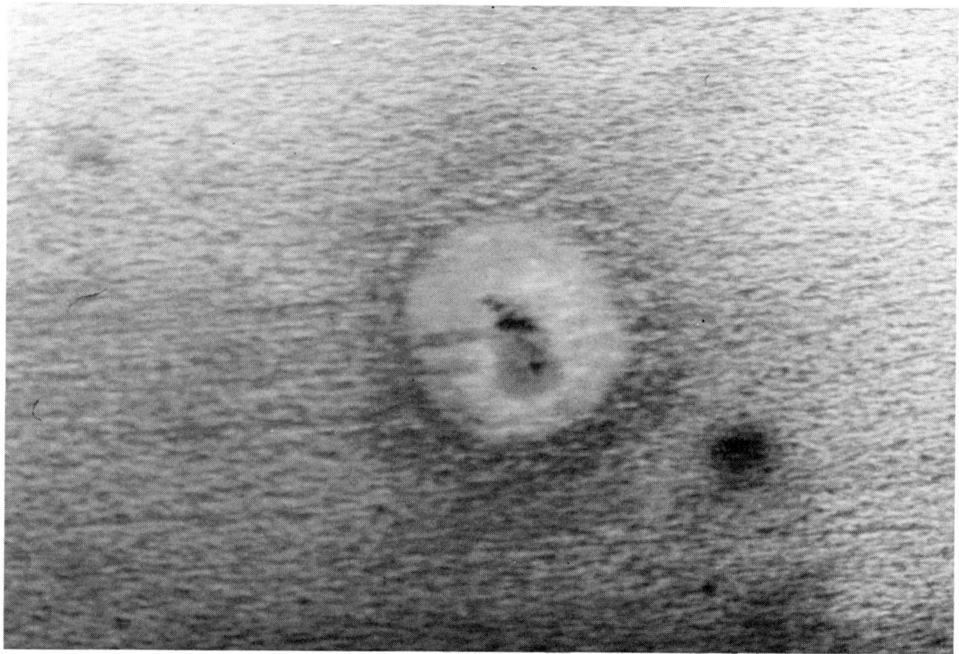

Figure 7. Photomicrograph of an imperfection on a Winchester-type magnetic disk.

cm^{-1} absorption band is absent in the imperfection spectrum. The features in the CH stretching region are different and new bands are observed in other regions as well. Strong bands are observed in the imperfection spectrum near 3400 and 1600 cm^{-1}. There are also similarities between the bulk coating spectrum and the imperfection spectrum. For example, the ester C=O and C-O stretching bands at 1740 and 1230 cm^{-1}, respectively, are observed in both spectra. The interpretation of these data is somewhat speculative; however, it appears that a couple of different mechanisms could be responsible for the imperfection. First, it is believed that the 3400 and 1600 cm^{-1} bands are due to water bound to an agglomeration of iron (III) oxide. Unfortunately, the iron (III) oxide absorptions are below 700 cm^{-1} and could not be observed utilizing a narrow-band MCT detector. The visible light microscopy supports the iron (III) oxide agglomeration thesis. Secondly, it appears that the polyurethane has degraded or an organic impurity was somehow introduced. This explanation is required to account for the presence of new bands in addition to the hydrated water bands. In summary, these data are consistent with a chemical

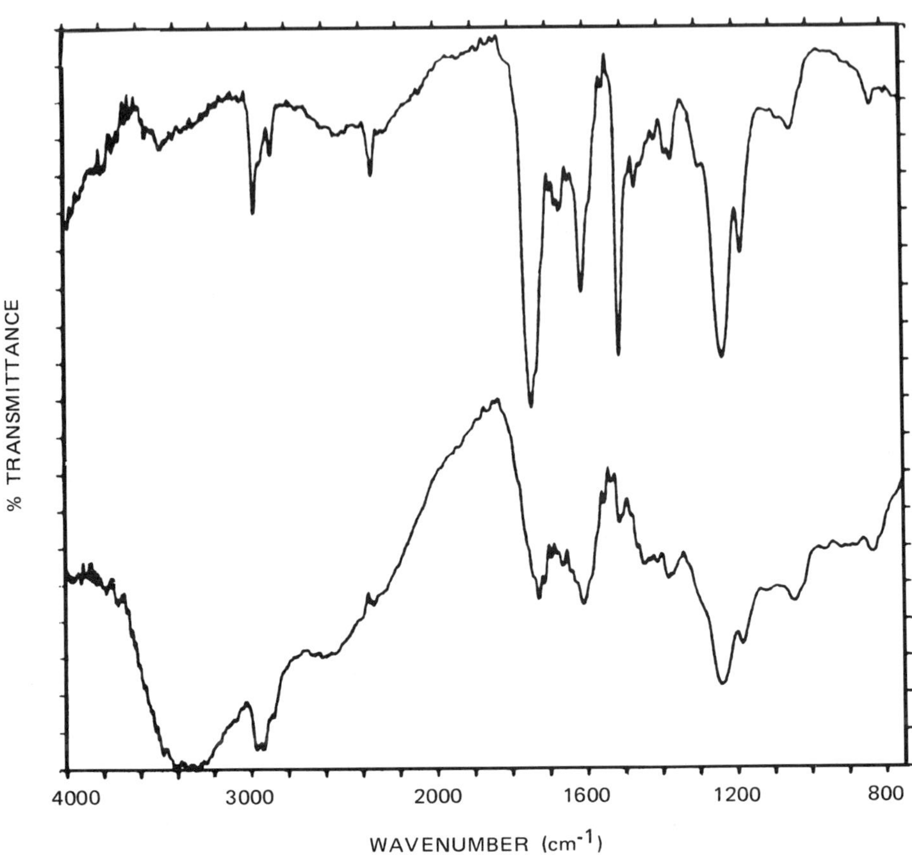

Figure 8. Top: Winchester disk matrix. Bottom: Imperfection area. Conditions: model 1800, IR-plan microscope, narrow-band MCT detector (immersed, .20mm X .20mm element), 8 cm^{-1} resolution, 128 scans.

dissimilarity between the imperfection area and bulk coating and point out the potential of infrared microspectroscopy to solve this difficult problem which appears to be insoluble by other infrared spectroscopic means.

Another application of reflectance infrared microspectroscopy is the analysis of lubricants on the edge(s) of razor blades. The analysis of these lubricants is important for manufacturing and quality control problems as well as competitive analyses. These lubricants prevent oxidation of the metal blade and also allow a smooth shave. Figure 9 is a photomicrograph of the edge of a razor blade. A remote image masking aperture was set to a slit width corresponding to 12 X 300 μm at the sample, and this slit was aligned with the

Figure 9. Photomicrograph of the edge of a razor blade.

edge of the blade. The reflectance spectrum in figure 10 was obtained. A hydrocarbon material is present as evidenced by the CH stretching and CH bending modes. Bands near 1230 and 1150 cm^{-1} are also observed. These bands are presumably due to a thin Teflon® coating on the edge of the blade.

Samples for transmission studies are typically measured neat with an infrared microscope rather than dispersing them in a nonabsorbing material such as KBr as is generally done for IR microsampling measurements with beam condensers. However, many neat samples have pathlengths which are too great and poor spectral representations are observed. In general a pathlength of 10-20 μm gives a good spectral representation (this is of course dependent on the sample). Different methods have been noted for flattening samples. These include the use of a probe, hydraulic or hand presses, or a cylindrical rod in contact with a hard surface. Each of these methods require subsequent removal of the sample for IR spectral measurement which could be very disadvantageous for small, unique samples. Diamond anvil cells have been suggested [7,8] as convenient accessories for flattening and holding samples for IR microspectroscopic measurements. Diamond is the hardest substance known

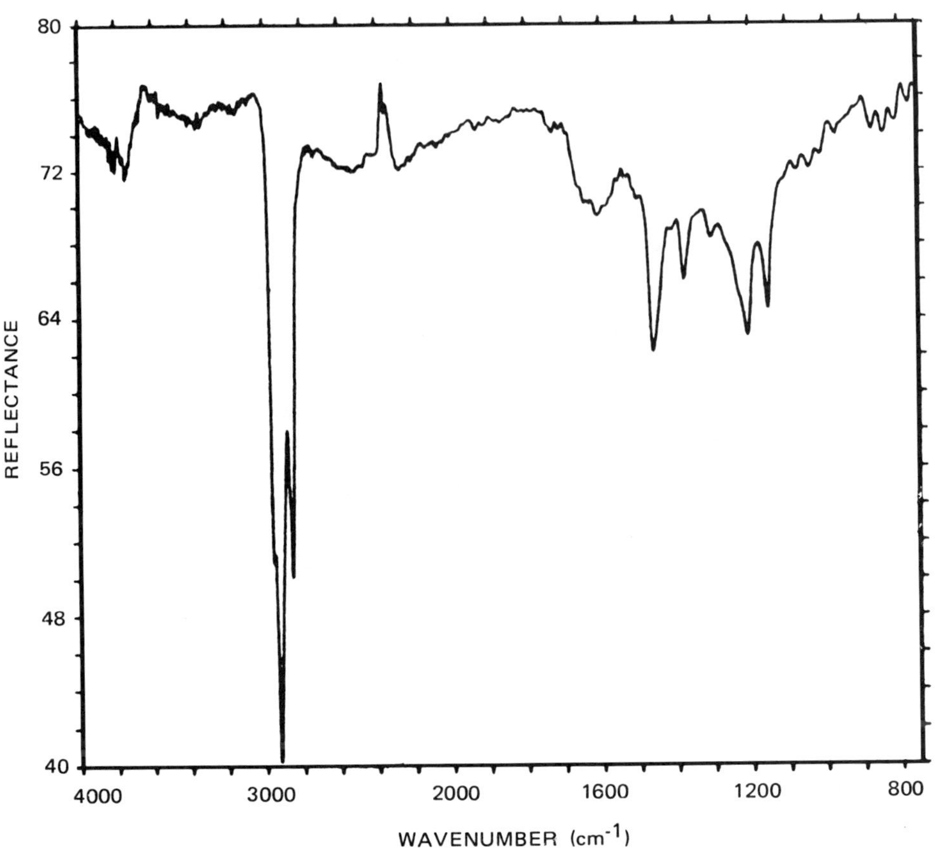

Figure 10. Reflectance spectrum of the edge of a razor blade. Conditions: model 1800, IR-plan microscope, narrow-band MCT detector (immersed, .20mm X .20mm element), 8 cm^{-1} resolution, 128 scans.

and is very transparent in the mid-IR region. Figure 11 is a photomicrograph of some material flattened in the miniature diamond anvil optical cell. A barely noticeable interference pattern is observed across the faces, an indication of good parallelism. Using this method, the parallelism of the diamond faces should be checked each time the cell is used. If the faces are not parallel, the diamond cell may be damaged when pressure is applied. Unfortunately, this high degree of parallelism can give rise to interference fringes in the spectra.

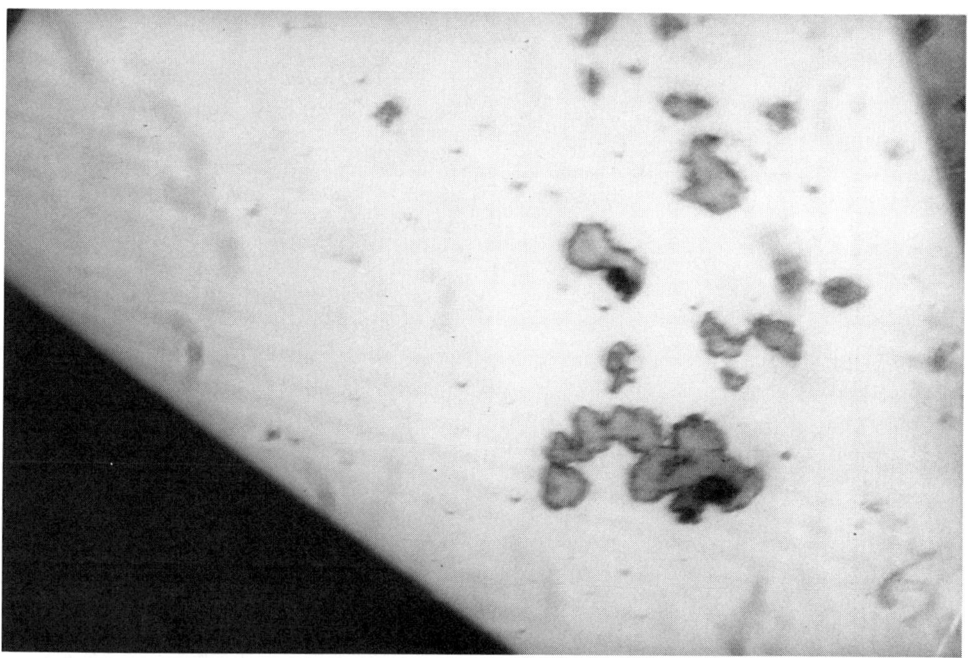

Figure 11. Photomicrograph of material held in the miniature diamond anvil optical cell.

The dark area in the bottom left corner of Figure 11 is an image viewed through the diamond cell of one of the blades of an aperture below the lOX condensing Cassegrainian objective of the IR-plan microscope. The lensing effect of the diamond cell has been corrected in the visible region by adjusting the separation between the primary and secondary mirrors of the Cassegrainian optics. Figure 12 is a spectrum of a substance identified as a carbonate, a common industrial contaminant. This spectrum was obtained using the miniature diamond anvil optical cell in conjunction with the IR-plan microscope. The sampled area was of approximately 30 μm dimensions. Carbonate is a strongly absorbing species; however, the observed spectrum is of good quality indicating a flattened sample.

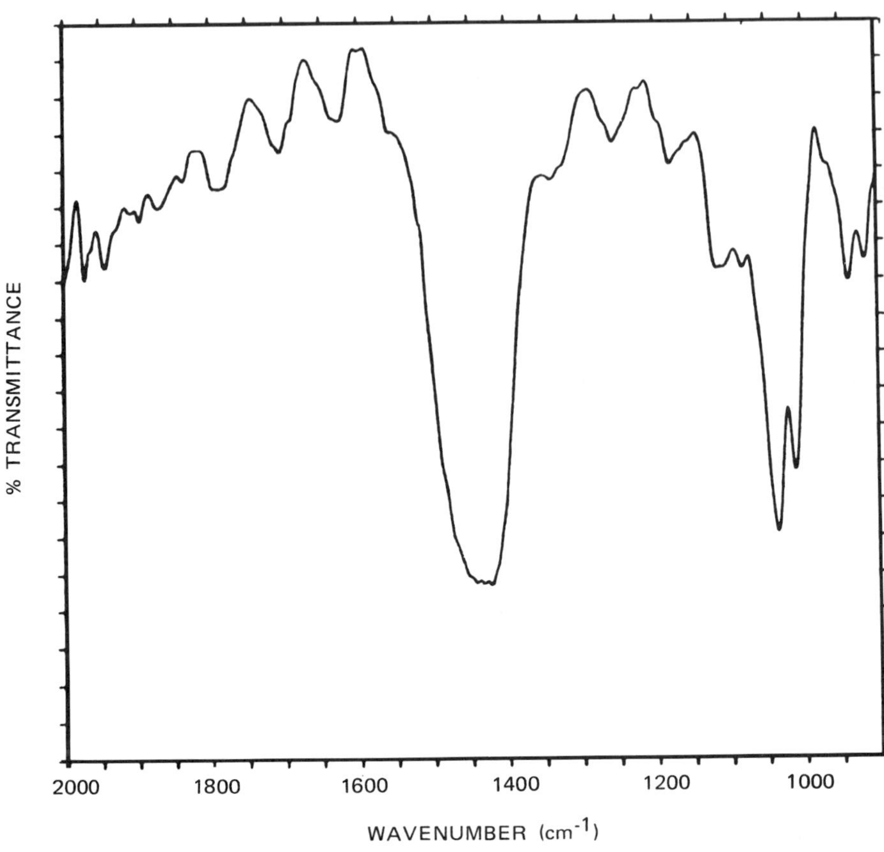

Figure 12. Spectrum obtained through the miniature diamond anvil optical cell. Conditions: model 1750, IR-plan microscope, microscope mounted narrow-band MCT detector (.25mm X .25mm element), 8 cm-1 resolution, 128 scans.

16.4 CONCLUSION

In this report attention was focussed on the application of infrared micro-spectroscopy to the solution of manufacturing and quality control problems in samples from various concerns. Current commercially available microscopes allow both transmission and reflection measurements, expanding the applicability of the technique. As demonstrated both particulate samples and small areas in larger matrices may be identified. The technique was applied to diverse samples including imperfections in polymers, a contaminant on a silicon wafer, and an imperfection on a magnetic disk. Reflection spectra were recorded for samples on opaque substrates.

REFERENCES

1. R. Barer, A. R. H. Cole, and H. W. Thompson, *Nature*, 163: 198 (1949).
2. V. J. Coates, A. Offner, and E. H. Siegler, Jr., *J. Opt. Soc.*, 43: 984 (1953).
3. 12th Federation of Analytical Chemistry and Spectroscopy Societies meeting, Philadelphia, PA (1985); Pittsburgh Conference, Atlantic City, NJ (1986); 13th Federation of Analytical Chemistry and Spectroscopy Societies meeting, St. Louis, MO (1986); 25th Eastern Analytical Symposium, New York, NY (1986).
4. R. G. Messerschmidt, *Paper 336*, Pittsburgh Conference, Atlantic City, NJ (1986).
5. D. B. Chase, *Appl. Spectrosc.*, 38: 491 (1984).
6. M. A. Harthcock, L. A. Lentz, B. L. Davis, and K. Krishnan, *Appl. Spectrosc.*, 40: 210 (1986).
7. P.L. Lang, J.E. Katon, D.W. Schiering, and J.F. O'Keefe, *Polym. Mater. Sci. Eng.*, 54: 381 (1986).
8. J. E. Katon, P. L. Lang, D. W. Schiering, and J. F. O'Keefe, *Instrumental and Sampling Factors in Infrared Microspectroscopy*, in *The Design, Sample Handling, and Applications of Infrared Microscopes* (P. B. Roush, Ed.), ASTM STP 949, American Society for Testing and Materials, Philadelphia, p.49 (1987).

17

Applications Of Molecular Microspectroscopy To Paper Chemistry

A. J. SOMMER, PATRICIA L. LANG, BRIAN S. MILLER, AND J.E. KATON *Molecular Microspectroscopy Laboratory, Department of Chemistry, Miami University, Oxford, Ohio*

17.1 INTRODUCTION

Although some fundamental work has been reported on Raman microspectroscopic studies of cellulose and lignin [1], there appears to be little or no work reported on applications of microspectroscopy to problems within the paper industry. The recent success experienced in our laboratory with the application of infrared microspectroscopy to studies of synthetic and natural fibers [2], has prompted us to investigate the application of this technique to paper chemistry. We report herein preliminary findings on the use of infrared microspectroscopy for the characterization of paper and paper products. Examples are also given of the application of this technique to practical problems encountered within the paper industry. To allow the reader a better understanding of the results presented, basic concepts and steps employed in paper manufacturing will be reviewed.

Four main operations employed in paper manufacturing are pulping, stock preparation, web formation and web modification. The first operation, pulping, involves the liberation of fibers from wood chips or recycled paper by breaking down the glue that binds the fibers together. In the case of wood chip stock, fibers are liberated by breaking down the natural glue, lignin. Pulping of secondary fibers (fibers from recycled paper) necessitates the removal of glues, inks, and other chemicals used in the manufacture of the original paper. After fibers have been liberated they can be bleached to further remove the lignin and are sent on to be used in stock preparation.

In stock preparation the blending of all materials needed to manufacture a crude paper takes place. During stock preparation the fibers are refined to yield a uniform size and to increase the number of hydroxyl groups per fiber. These refinements increase the strength of the finished paper product by increasing hydrogen bonding between fibers. The fibers are then diluted with water and chemicals to approximately 1% yielding a mixture called the "furnish". A tank, or headbox, delivers the furnish to a conveyorized screen where the liquid portion of the furnish is removed. The solid portion remaining on the conveyor forms a crude paper, called the web, which is then pressed, dried and modified.

At this point paper manufacturer's utilize a number of chemicals to impart desired properties to the finished paper product. Some common properties to be improved through web modification are water resistance, opacity, and printability. Web modification by the addition of sizing agents and pigments may take place during web formation (internal sizing) or after the web has formed (surface sizing) [3]. Table 1 lists some common additives along with their specific task and chemical make up.

Table 1. Common Additives Used in the Paper Industry

Additive/ Application	Property Enhanced	Chemical Composition
Internal size	water resistance fiber bonding	rosin, alum, starch, animal glues, animal gums, vegetable gums and melamine formaldehyde
Surface size	water resistance fiber bonding	starch, polyvinyl alcohol, guar, and animal gums, waxes
Fillers (internal and surface)	opacity printability	kaolin, $CaCO_3$, and TiO_2

Table 1. (cont.)

Additive Application	Property Enhanced	Chemical Composition
Pigmented Coatings	opacity printability water resistance	Pigments: kaolin, $CaCO_3$,TiO_2 alum, lime, polystyrene Adhesives: starch, casein, and latex emulsions of acrylics, polyvinyl acetates and styrene/butadiene

Because there are a vast number of paper products produced in today's technology, many different chemicals are used to enhance the properties of papers manufactured for specific applications. However, those additives listed in Table 1 form the basis of many of these paper products. For specific refinements and a more detailed analysis of paper manufacturing the reader is referred to an excellent account of technology presently employed in the paper industry [3].

17.2 EXPERIMENTAL

Micro infrared spectra were obtained with the aid of an Analect (Irvine, CA) AQS-20M micro infrared system consisting of a fXA-6260 interferometer and a fXA-515 microscope module. A Bausch and Lomb (Rochester, NY) stereo microscope was used to observe samples during sample collection and preparation. Microsamples of interest were removed from the parent matrix with a fine pointed probe and transferred to a 6 mm diameter KCl window. Thin sample sections were obtained by flattening the sample on a glass slide prior to transfer to the salt plate. Spectra were collected by signal-averaging 128 scans. Representative spectra of macro samples were confirmed by micro-sampling at least three different areas of the sample.

17.3 RESULTS AND DISCUSSION

Preliminary studies of the application of infrared microspectroscopy to paper chemistry involved the characterization of some of the raw materials used in paper manufacturing. Figure 1 shows the infrared spectra of a hardwood fiber and a softwood fiber. Both spectra have infrared absorptions characteristic of lignin and cellulose; however on comparison, the spectrum of the softwood fiber exhibits additional absorptions at 1694 and 1269 cm^{-1}. These bands can be assigned to rosin present in the softwood fiber. Softwood, or coniferous, trees have a higher rosin content than do hardwood trees.

Bleaching of pulped fibers serves to remove residual lignin left by the pulping process and brightens the fibers, giving the finished paper product longer permanence. Lignin remaining in paper will tend to oxidize and ultimately yellow the paper. Spectra shown in Figure 2 are representative of papers manufactured from fibers with and without residual lignin. Absorption bands characteristic of lignin are observed in the top spectrum at 1736, 1656,

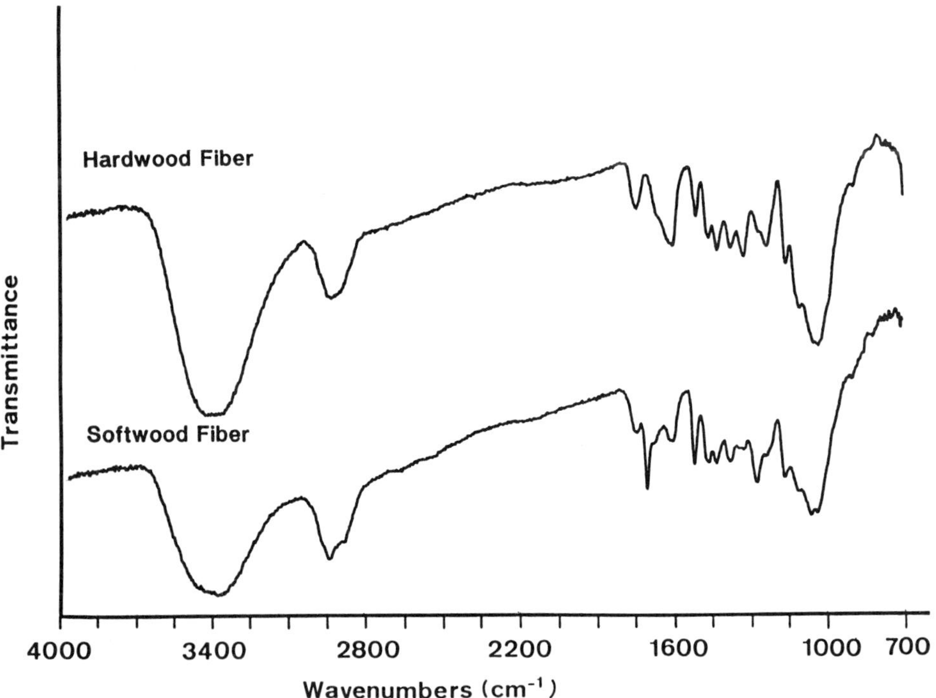

Figure 1. Infrared spectra of hardwood and softwood fibers.

1600, 1503, and 1235 cm^{-1}. These absorptions are absent in the bottom spectrum. Absorptions present in the bottom spectrum are characteristic of pure cellulose.

Infrared spectra shown in Figure 2 are spectra obtained from a paper with no additives and the same paper exposed to a corona discharge in oxygen for 30 seconds. Treatment of cellulose strips in a corona discharge has greatly improved bond strength when the strips were wetted, pressed together and then dried [4]. At present no chemical explanation for this phenomenon is available, however, evidence suggests that the discharge decomposes the surface of the paper to yield a material similar to hemicellulose. Hemicellulose bonds better than cellulose under these conditions, indicating that the material formed produces a surface adhesive responsible for the increased bond strength.
As discussed previously the only difference between these two spectra is the presence and/or absence of lignin. Clearly the discharge has removed the lignin, presumably by some sort of degradation mechanism, and has not

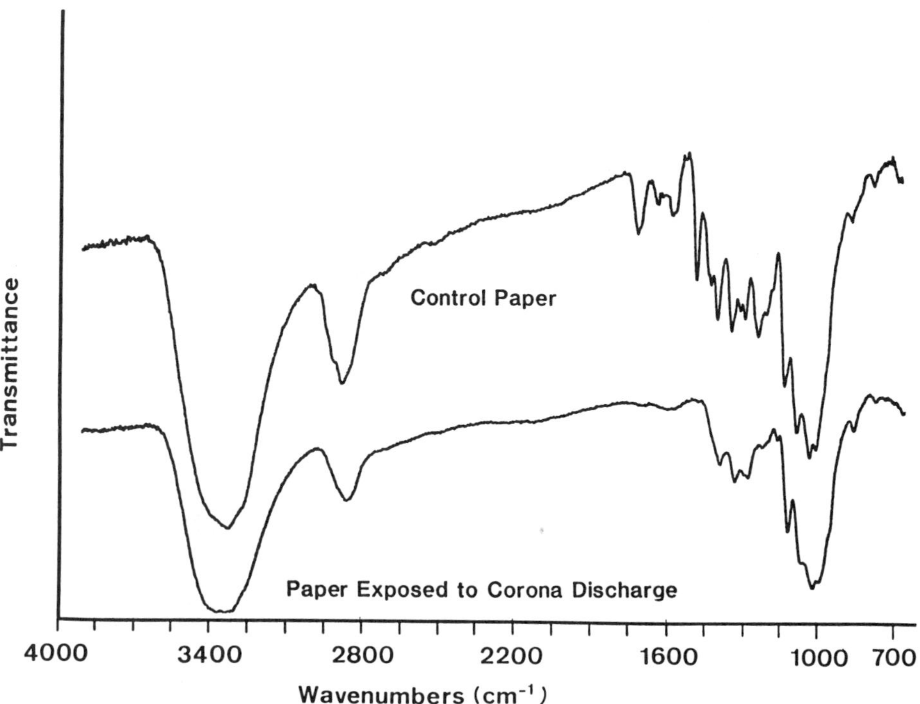

Figure 2. Infrared spectra of a control paper and the same paper exposed to a corona discharge.

detectably altered the cellulose. The paper involved in this experiment was newsprint paper, which contains a large amount of lignin on the surface of the cellulose fibers. The expected oxidation of cellulose was not observed, perhaps because of insufficient time of exposure to the discharge. The fibers investigated were surface fibers. The lack of reference spectra for well characterized hemicellulose and its similarity to cellulose make it difficult to follow any spectral changes associated with the formation of this material.

Water resistance, opacity and printability are all improved by the addition of a surface or internal sizing agent and pigments. Spectra obtained from a microtomed cross-section of a finished paper product and cellulose are presented in Figure 3. In addition to infrared absorptions characteristic of cellulose, the spectrum of the finished paper product exhibits prominent bands located at 3696, 3619, and 1726 cm^{-1}. The first two absorptions are characteristic of kaolin which is a common filler or pigment used in the paper industry. The latter band is indicative of an ester functionality, however, interfering absorptions of the cellulose inhibit a clear identification. Figure 4

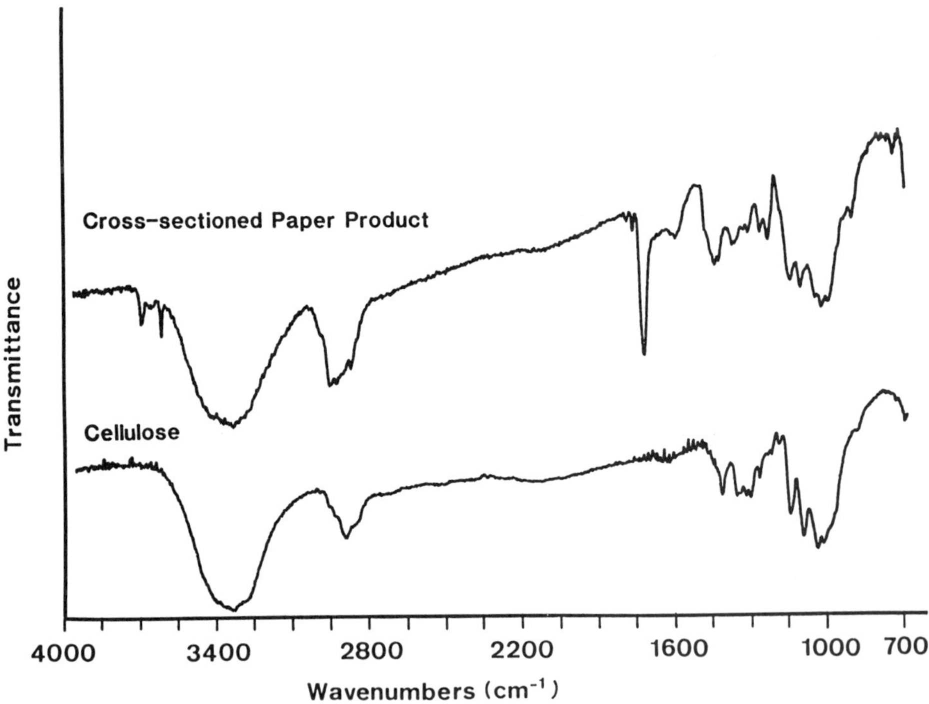

Figure 3. Infrared spectra of a cross-sectioned paper product and cellulose.

shows the spectrum of the finished paper with cellulose spectrally subtracted. Absorptions enhanced by the subtraction make it possible to identify the ester as a polymethacrylate commonly used as an adhesive in surface sizings. Unfortunately, in this case, polymethylmethacrylate was used as the embedding medium, and therefore one cannot be certain that the source of this material was the original fiber.

Samples discussed thus far could have been analyzed using conventional macrosampling techniques, but only with a great deal of effort in sample collection and preparation. The remaining data to be discussed demonstrate not only the ability of infrared microspectroscopy to be used as a characterization tool for macrosamples, but also the intended use of the technique, particulate analysis.

One topic favorite to everyone is that of money. Our interests in this topic, however, were solely scientific in nature. New one dollar bills were obtained from a local bank, and the different inks on each bill were sampled to

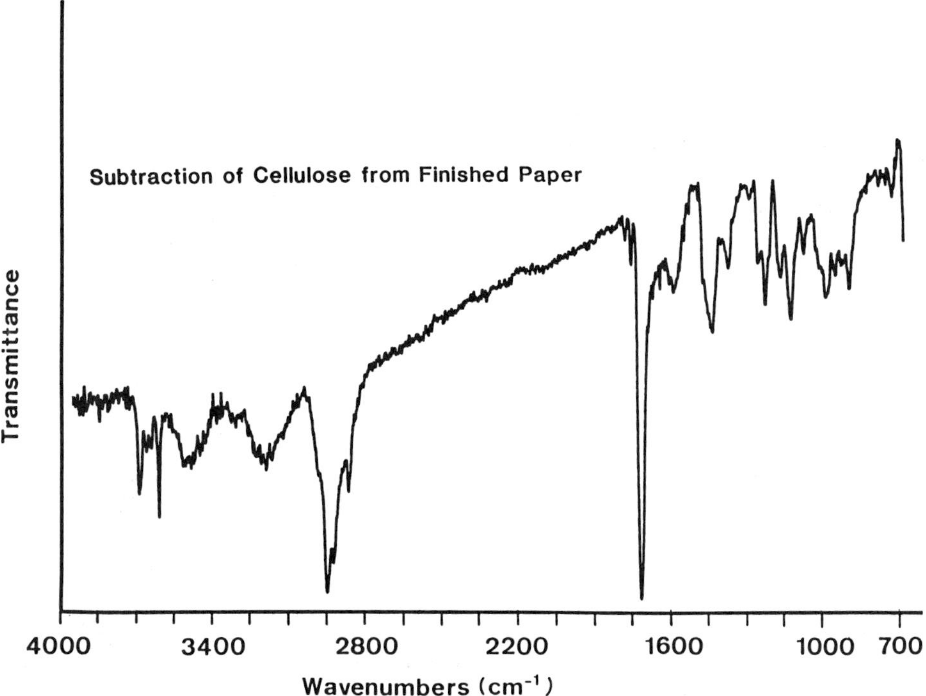

Figure 4. Spectrum of finished paper product with cellulose spectrally subtracted.

see if any differences could be established. Figure 5 shows the infrared spectrum of ink scraped from the light green serial numbers. Absorption features in this spectrum are characteristic of kaolin and an adhesive based on an acrylic. Figure 6 shows infrared spectra of the black and green inks which comprise the major printed material on this currency. The spectra demonstrate that there are slight differences between these two inks and that their composition is mainly calcium carbonate and an acrylic adhesive. Carbonate bands are observed 1422 and 873 cm^{-1} while those of the acrylic are observed 2955, 2919, 2852, and 1725 cm^{-1}. Two absorptions which remain unaccounted for are observed at 2512 and 1795 cm^{-1}. These bands may arise from the pigment used, however, in this laboratory infrared studies of inks on paper have, to this date, produced only one spectrum in which a colored pigment has been identified. This spectrum is shown in Figure 7 and was obtained from green ink scraped off a piece of graph paper. The spectrum exhibits absorption bands characteristic of cellulose, kaolin, and lignin. An additional band located at 2100 cm^{-1} has been assigned to a mixed iron cyanate/lead chromate pigment

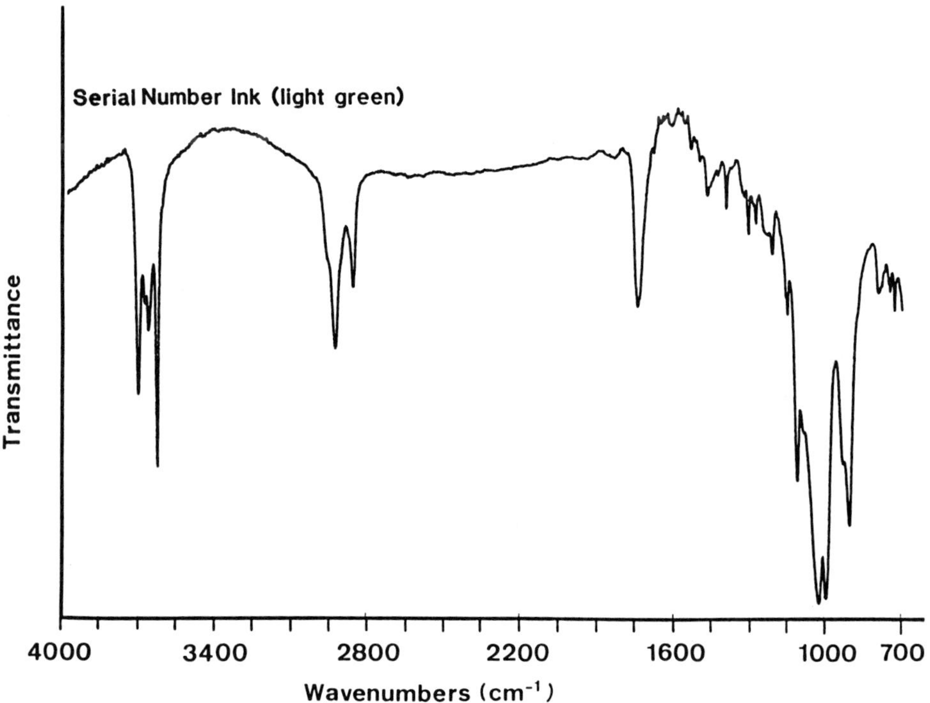

Figure 5. Infrared spectrum of green ink scraped from a dollar bill.

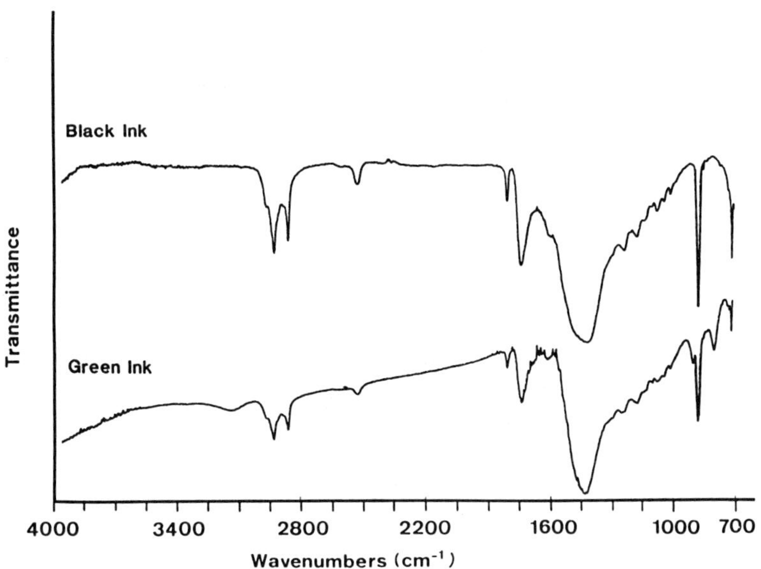

Figure 6. Infrared spectra of green and black inks removed from a dollar bill.

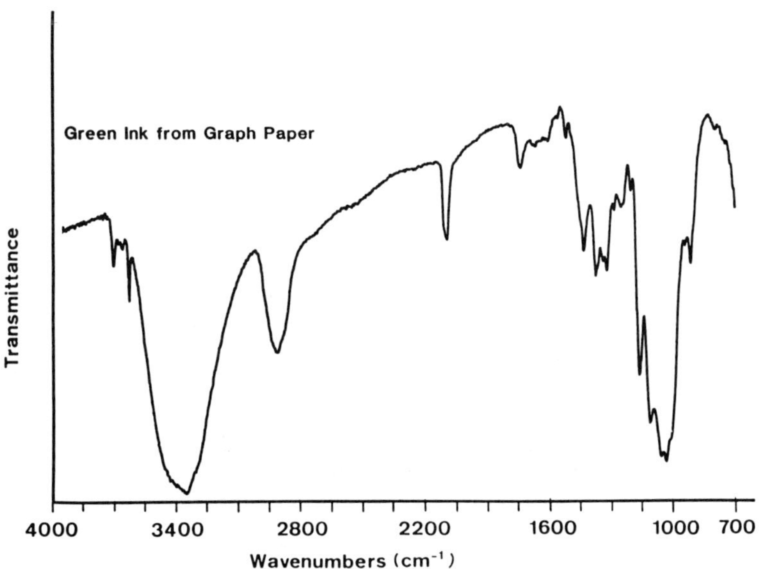

Figure 7. Infrared spectrum of a green ink removed from graph paper.

known as chrome green [5]. The difficulties encountered in observing pigments in inks may come from the use of inorganic pigments or because the amount of pigment used may be below the detection limit of the technique. An interesting observation to end this section came from the infrared spectrum (Figure 8) of colored fibers removed from the dollar bills. These fibers were woven into the cellulose matrix of the bills at random locations. The fibers could easily be identified as nylon and are intentionally placed in the bills to prevent duplication and/or counterfeiting.

The success of recycling paper depends upon the removal of materials used to manufacture the original paper. Inclusion of polymers and plastics in the wastepaper may make it unusable or cause considerable problems during processing [3]. One aspect to which infrared microspectroscopy may be useful is the identification of these particulates in recycled paper. Should the manufacturer determine the source of these contaminants he or she may take the necessary steps to eliminate it from the process. Two recycled paper products which we have studied are the inner liner of cereal boxes and paper towels. Observation of these products under a stereo microscope show that they contain

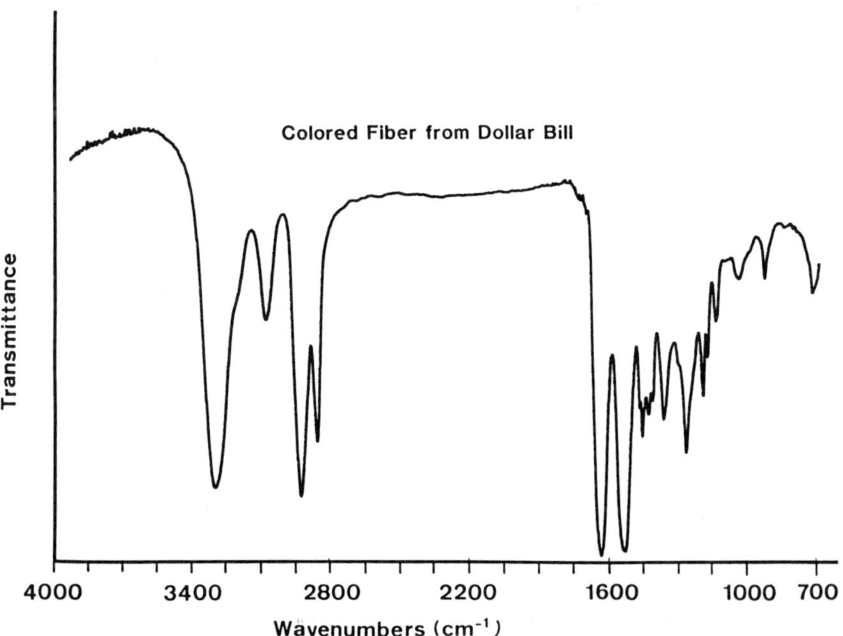

Figure 8. Infrared spectrum of a colored fiber removed from a dollar bill.

a large amount of particulate contaminants. Two categories into which these contaminants fall are brown granular residues and polymer particulates.

Figure 9 shows infrared spectra of brown residues removed from the inner liner of the cereal box and the paper towel. The spectral features in these spectra are characteristic of cellulose containing lignin. The presence of lignin indicates that the wastepaper used to manufacture these products was probably manufactured from unbleached pulp. Polymer particulates taken from the paper towel produced the spectra shown in Figures 10 and 11. The spectrum shown in Figure 10 matched that of an ethacrylate/styrene copolymer. This type of polymer is used as an adhesive in pigmented coating systems. Infrared absorptions observed in the spectrum of Figure 11 are characteristic of kaolin, calcium carbonate, cellulose, and an acrylic polymer. The identity of the acrylic, unfortunately, could not be determined. Finally, the spectrum shown in Figure 12 was obtained from a polymer removed from the inner liner of a cereal box. This polymer was identified as cellulose propionate, which is commonly used as a packaging material.

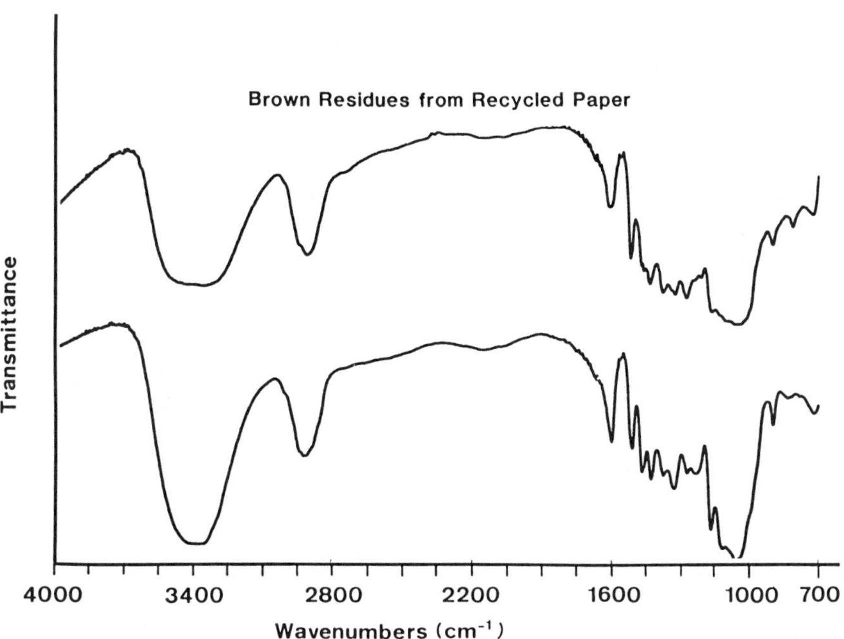

Figure 9. Infrared spectra of brown residues removed from a paper towel and the inside liner of a cereal box.

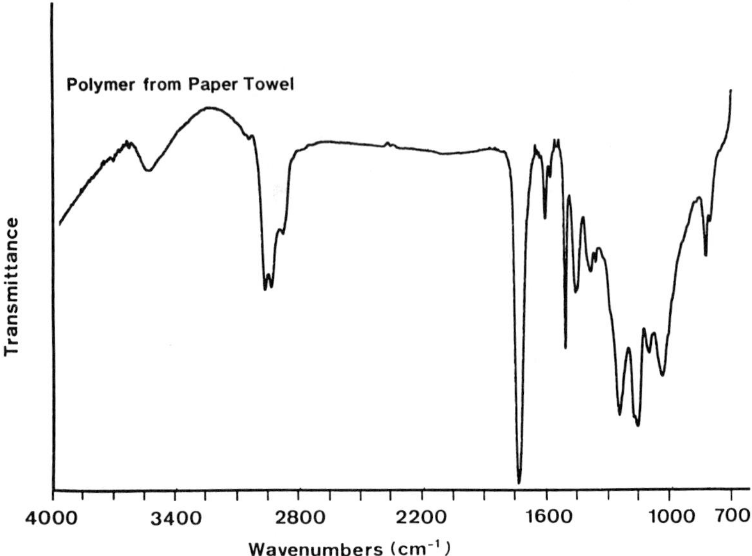

Figure 10. Infrared spectrum of a polymer removed from a paper towel (ethacrylate/styrene copolymer).

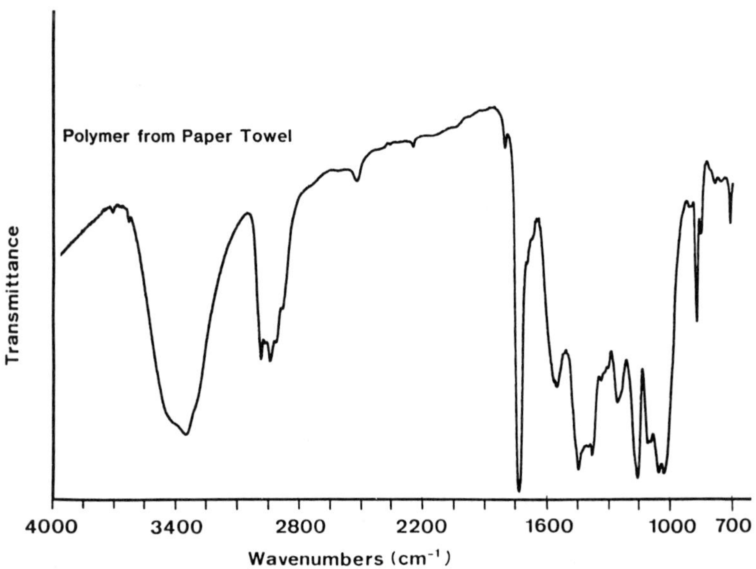

Figure 11. Infrared spectrum of a polymer removed from a paper towel.

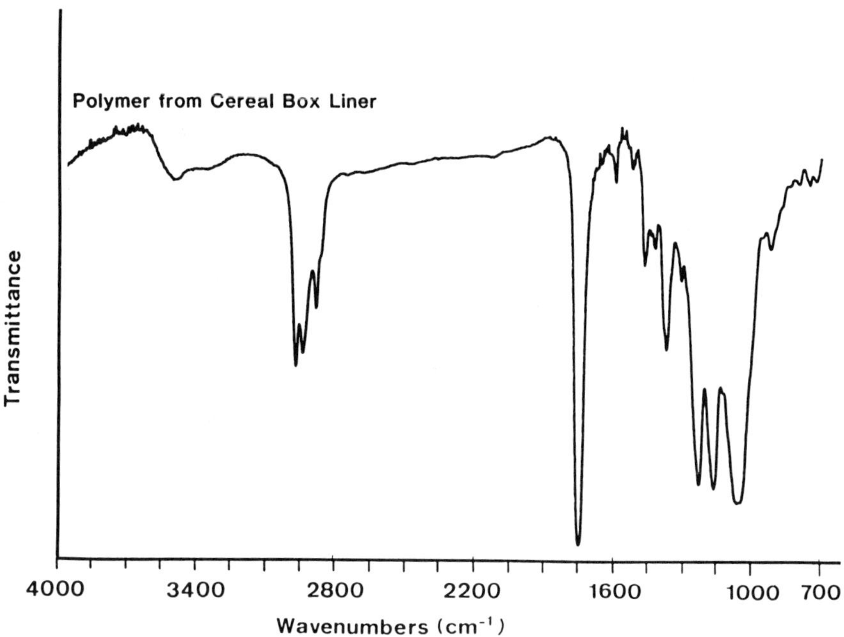

Figure 12. Infrared spectrum of a polymer removed from the inside liner of a cereal box.

17.4 CONCLUSIONS

The results presented in this investigation clearly demonstrate that infrared microspectroscopy can be a valuable investigative tool when applied to paper chemistry. While some of the examples cited could have been studied via conventional infrared techniques, our experience with infrared microspectroscopy has been that analysis time is greatly reduced solely from that fact that analysis requires little sample with minimal preparation. Care must be taken however to ensure that spectra from multiple sampling are identical when this technique is used to characterize macrosamples.

REFERENCES

1. R. H. Atalla and U. P. Agarwal, *Science,* 227: 636 (1985).
2. J. K. Katon, P. L. Lang, D. W. Schiering and J. F. Okeefe, *Paper 21, 12th Annual Meeting* FACSS (1985).

3. James E. Kline, *Paper and Paperboard: Manufacturing and Converting Fundamentals,* Miller Freeman Publications, San Francisco (1982).
4. D. A. I. Goring, *Pulp and Paper Magazine of Canada,* 68(8): T372 (1967).
5. Chicago Society for Coatings Technology, *An Infrared Spectroscopy Atlas for the Coatings Industry,* Federation of Societies for Coatings Technology, Philadelphia, PA (1980).

18

Infrared Microscopic Reflectance Studies of Used Test Specimens

DAVID L. WOOTON AND DENNIS W. HUGHES *Ethyl Petroleum Additive Division, St. Louis, Missouri*

18.1 INTRODUCTION

The interaction of chemical compounds with a metal surface has been the focus of attention for decades. For those interested in developing lubrication oils, this information is of vital importance. The lubrication oil used to protect the metal surface can decompose on the surface to form the protective layer or the decomposition can form deposits that can damage the metal surface. To develop an effective product that can offer maximum surface protection, it is imperative that the surface deposits formed can be structurally identified.

In the actual testing of oils, the rubbing of metal surfaces can cause many types of chemical interactions to occur simultaneously. By means of a microscopic study, one can identify wear/deposit patterns on the metal surface. These different patterns contain different types of chemical components. With the recent developments in microscopic infrared spectroscopy, it is possible to identify the chemical composition on the physical surfaces of the metal.

With infrared spectra one can identify geometric and structural data of a molecule. If material is present on a metal surface, the application of infrared reflectance spectroscopy can allow the identification of compounds that exist on the metal surface. With this information the mechanism of surface interaction with the chemical species can be identified and thus can lead to the development of mechanistic wear theories. With this information it is potentially possible to help develop improved antiwear chemical compounds.

In the study of wear, what has occurred near the wear scar can help explain the mechanisms of wear. Wear scars are typically small (3 mm and smaller). Traditional reflectance infrared spectroscopic techniques can not be used to study these small areas. With the advent of microscopic equipment for infrared spectroscopy, techniques have been developed to study wear areas. This chapter was directed toward the development of the technique of reflectance microspectroscopic infrared of worn metal surfaces with very low surface coverage.

18.2 EXPERIMENTAL

The samples used in this study were obtained from two sources and were generated where two metal surfaces rub together with a lubricant providing a protective film. The mechanical rubbing that occurred was generated from two known wear testing instruments, the Timken Wear Tester [1] and the Four-Ball Tester [2]. These instruments differ by the speciman configuration and the load applied to the suface (Table 1).

Table 1. Comparison of Test Mechanics

	Timken	Four-Ball
ASTM Number	D2782	D2783
Test Specimen Configuration	Ring on Block	Ball on Ball
Test Geometry	Line Contact	Tetrahedron (Point contact)
Rotation Speed	800 rpm	1800 rpm
Scar Configration	Rectangular	Circular
Initial Temperature	100° F	65°-85° F
Test Length	10 min	10 sec
Actual Load on Specimen	10X Indicated	As Indicated
Lubricant Delivery System	Recirculation (1 gallon)	Immersed (10 grams)

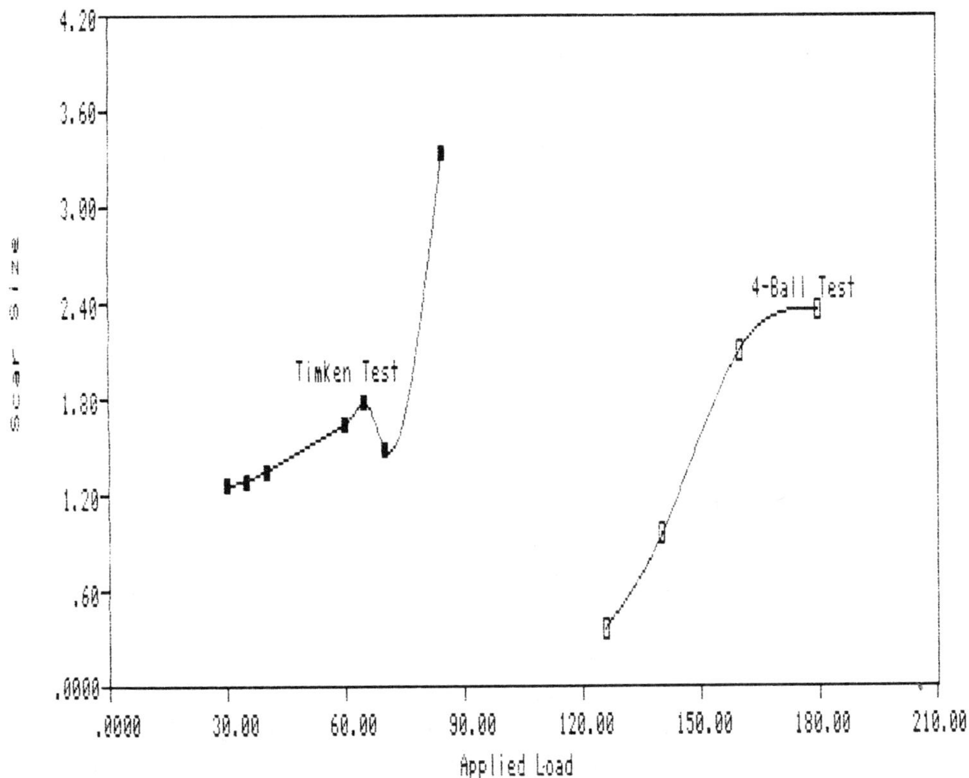

Figure 1. Test results - applied load versus scar size.

With the Timken Wear Tester, a ring wearing on a block, generates a line contact scar. The other wear sample, from a Four-Ball Wear Tester, has a rotating ball worn against three fixed balls in a tetrahedron configuration, forming a point contact and a circular scar. The scar size in both testers is dependent upon the applied load as seen in Figure 1.

The lubricant used to protect the wearing surfaces contained a hydrocarbon of C20 to C30 molecular size (95%). The remainder of this oil consisted of t-butyl polysulfide, alkyl amine and alkyl acid phosphate. Each of the additives used are for the protection or modification of the metal surfaces.

The infrared spectral data presented in this paper was obtained using a Perkin-Elmer 1800 (Ridgefield, CT) Fourier Transform Infrared Spectrophotometer. The microspectroscopic data was obtained using a Spectra-Tech (Stamford, CT) Surface-Scope accessory (a 60-80° grazing angle infrared microscope), and the macrospectroscopic data was obtained using a Spectra-Tech FT-80 (a fixed 80° grazing angle reflectance accessory).

The spectra were obtained using the single ratio technique. The background spectrum was from a cleaned metal surface and will be addressed in more detail later in the text. Scanning accumulations were added until typically 4000 to 6000 scans were obtained per spectrum. The atmosphere around the sample was dry nitrogen (since atmospheric carbon dioxide and water vapor can absorb more than the sample being analyzed). Some of the spectra presented were computer enhanced by smoothing and/or flattening to improve interpretation.

18.3 RESULTS AND DISCUSSION

18.3.1 Sample/Background Preparation

The preparation of the experimental sample and the background sample was found to be one of the most important factors effecting the analyses, since the surfaces being studied have very little sample on them. It was estimated that there was less than a monolayer of sample on the surface. Therefore, the presence of impurities, such as surface oil, can mask the sample being studied.

Preparation included washing the sample with pentane in an ultrasonic cleaner to remove the surface oil and the physically adhered materials. The chemically bonded compounds should remain on the surface. Exhaustive washing of the sample and the background specimens was a must. Figure 2 shows an example where the background sample was not washed sufficiently. The presence of a rust inhibitor, a carboxylic acid, on the background specimen can be seen as negative peaks. Further pentane washing of the reference ball eliminated this problem.

With some samples, not only is proper cleaning of both sample and reference needed, but both samples also needto see the same environmental conditions (Figure 3). The surfaces from the wearing experiment can have considerable structural changes caused by the loading on its surface. These structural effects can change the sample with respect to the background sample enough to obscure the reflectance infrared spectrum. In Figure 3, the bottom spectrum had a cleaned untreated surface as the reference background spectrum. The top spectrum had a reference background spectrum from a sample that was treated under the same loading wear conditions as the sample being studied; however, this sample did not have the antiwear additives in the oil. Two significant differences can be seen between these samples. First is the hydrocarbon observed in the bottom spectrum (3000 cm^{-1}) and second the surface effects from iron oxide (600 cm^{-1}) changes between the samples.

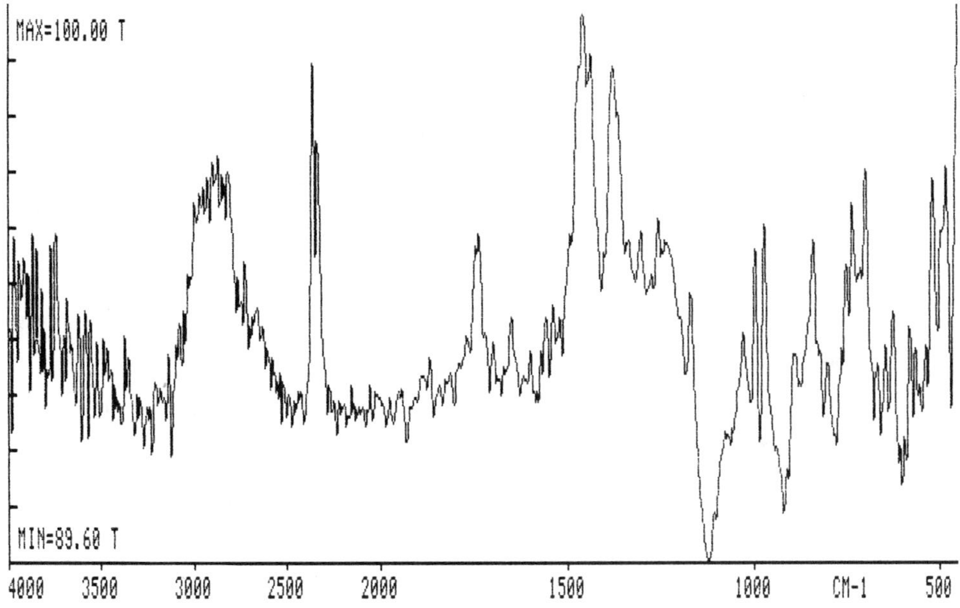

Figure 2. Effects of reference surface, 4-ball referenced against unused and solvent washed ball.

18.3.2 Instrumental Variables

The study of surfaces by reflectance infrared spectroscopy has numerous instrumental parameters that can effect the results. From previous results [3], the technique of spectral acquisition of single ratio was found to be the best method. With this method, the background interferogram was subtracted from the sample interferogram. The resulting interferogram was then transformed to produce the sample spectrum. This referencing technique allows the removal of the many instrumental spectral anomalies, including the affect of the infrared beam reflecting off many mirrors. It should also be noted that improper alignment of a reflectance microscope can impart spectral fringes that will overshadow the sample spectrum.

In reflectance infrared spectroscopy the angle which the incident light beam impinges onto the sample surface affects the result. This incident angle effects the pathlength of the deposit on the sample surface. With very thin coatings on a surface, the high angles yield increased sensitivity [4]. It has been

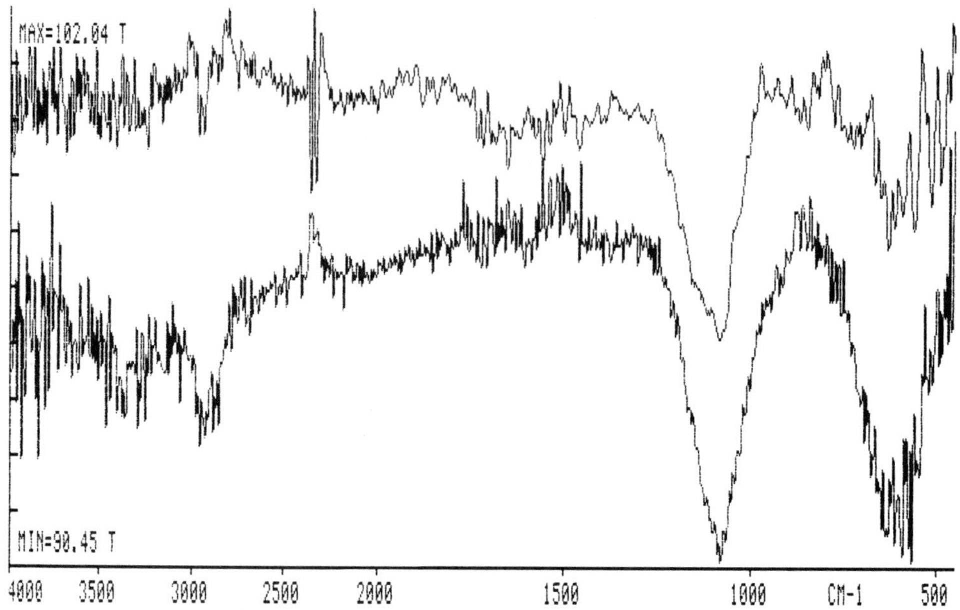

Figure 3. Timken effects of background sample on the 50 lb timken sample.
Top spectrum: 50 lb specimen with oil as reference sample.
Bottom spectrum: Cleaned, flat surface as reference sample.

determined from previous work that 85° is the optimum incident angle for this type of sample [4]. To approach this angle, the reflectance microscope used in this work was designed to have a 60-80° incidence angle.

The addition of a wire grid polarizer at 90° in the infrared beam has been a classic method of improving measurement sensitivity. Theoretically, this polarizer eliminates the perpendicular component of the surface vector which does not interact with the active dipole on the surface. The effects of a polarizer was studied using the reflectance infrared microscope. Comparisons of the same sample with the same number of accumulations, with and without a polarizer, showed no increased sensitivity with the polarizer. Due to the configuration of the microscope, the polarizer was placed before the Cassegrainian objective. This placement can allow a redistribution of the light polarization by the mirrors in the accessory before impinging on the sample. The polarizer on a reflectance microscope can not be used as the incident energy has radial symmetry which will scramble the polarization.

The detector of an infrared spectrophotometer can greatly effect the sensitivity of the sample. On the spectrophotometer used in this study, both a MCT (mercury/cadmium/tellurium) and a DTGS (deuterated triglycine sulfate) detector were available. As shown by Figure 4, the MCT detector yields significantly better sensitivity then the DTGS detector. It was necessary to run 1600 versus 2400 scans to obtain equal signal-to-noise ratios. Even though the sensitivity between the two detectors was different, the spectra generated did not differ. This says the spectrum was not a detector artifact but was independent of the detector. The MCT detector had a 600 cm^{-1} spectral cut-off, below which valuable data for this work was located. For this reason, the detector used in the rest of this work was the DTGS detector.

The spectral data presented thus far was obtained from the Surface-Scope, an infrared microscope. The surface area being studied consisted of a 500-micrometer diameter spot. In most cases, the position on the sample being studied was the center of the scar; however, the ability to tour the sample and compare various sections is available. With a macro spectral accessory, such as the FT-80, this option is not available. Spectral comparisons, as shown in Figure 5, demonstrate the differences between the microscope accessory and the

Figure 4. Effects of detectors - DTGS vs. MCT.
Top spectrum: DTGS (2400 scans).
Bottom spectrum: MCT (1600 scans).

macroreflectance accessories. The sample studied in Figure 5 was a Timken block at 50 lbs loading. This was the one sample configuration large enough to be studied by the macroreflectance accessory. The microscope spectrum was taken inside the wear scar, while the macroreflectance spectrum included the entire block face. It can be noted that the spectrum of the entire surface shows additional types of materials, possibly from around the scar.

18.3.3 Sample Condition

The next topic of discussion is the sample condition. This includes the surface roughness of the sample and the effect of the test conditions on the sample. The most often questioned parameter of a reflectance sample has been the surface roughness and the effect on the spectrum. From this study it has been shown that surface roughness effects the amount of energy reflected to the detector; however, this did not impair the ability to obtain a spectrum unless very large surface aberrations were observed. Spectral changes caused by surface roughness were studied by using the Timken block sample and changing its reflective configuration. Due to the Timken test conditions, that is rubbing on a block, the scar produced is rectangular in shape. If the angle of incidence of the infrared beam is perpendicular to this rectangular scar, the surface would be

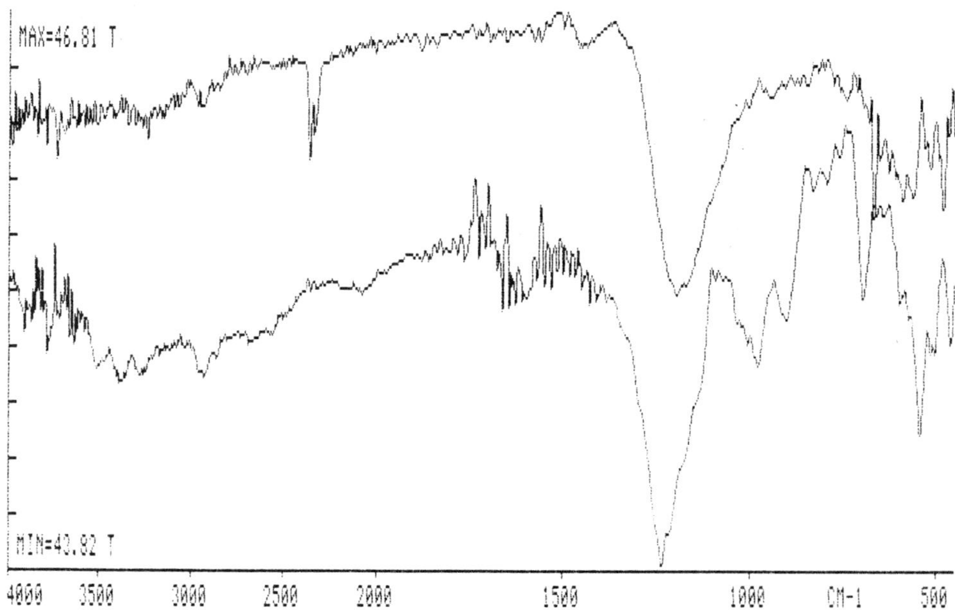

Figure 5. Differences between micro and macro infrared.
Top spectrum: Using Microscope reflectance.
Bottom spectrum: Using Macroreflectance.

relatively smooth. This is the direction of rubbing that caused the scar. If the incidence angle is parallel to the rectangular scar, the beam is reflecting across the roughness of the wear pattern. This direction change would cause an effective increase in roughness on the same sample. By comparing these two spectra, it can be determined that the sample roughness does not change the spectra. This concept opens the ability to compare two samples that were produced under different wearing conditions and having different surface roughness.

One of the variables on wear testing equipment is the load applied to the sample. As the load is increased on the sample the scar size and the roughness are increased. Since there was no spectral changes with a variation in surface roughness, spectral features can be compared with different loads. The loads on a Timken block did not greatly change the resulting spectra from inside the scar, Figure 6. The spectral changes in this figure were a shift of the peak maxima, a 15 cm^{-1} shift seen between the 30 and 40 lb loads and an additional 23 cm^{-1} shift between 40 and 60 lb loads. In addition, an absorption intensity increase was observed in the P-O-C stretch at 930 cm^{-1} as loading was increased. These changes indicate more oxidative deposition of the sulfur source and more involvement by the phosphorous source as the Timken load increased.

Figure 6. Effects of loading on Timken Block Spectra.
Top spectrum: 30 lb loading.
Middle spectrum: 40 lb loading.
Bottom spectrum: 60 lb loading.

The spectral effects of load changes on a Four-Ball Wear specimens were considerably greater, Figure 7. As the loading was increased less material was seen on the surface. Significant intensity changes were seen in the sulfur source peak, 1150 cm^{-1}, as the loading was increased from 126 to 160 kg. In this spectrum some of the hydrocarbon C-H stretchings were still observed, as well as the phosphorous source at 1000 cm^{-1}. At greater weights, that is 180 kg, most of the surface components were absent.

The sulfur source was added to the lubricant as an antiweld component. At 180kg load, there was no sulfur source observed on the surface. The result was that this additive was not functioning at this load. The high wear observed on this sample might have been from oil starvation during the wear, or the additive not being replenished fast enough at this load.

The analyses of a Timken block surface by a macroreflectance accessory showed that differences around the block surface exist. These differences can

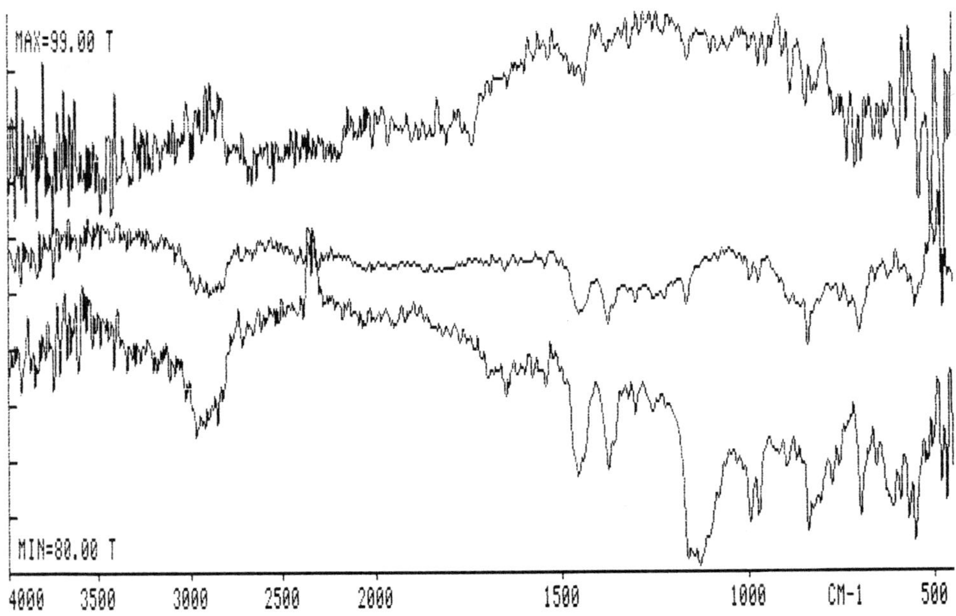

Figure 7. Effects of loading on Four-Ball Spectra.
Top spectrum: 180 kg loading.
Middle spectrum: 160 kg loading.

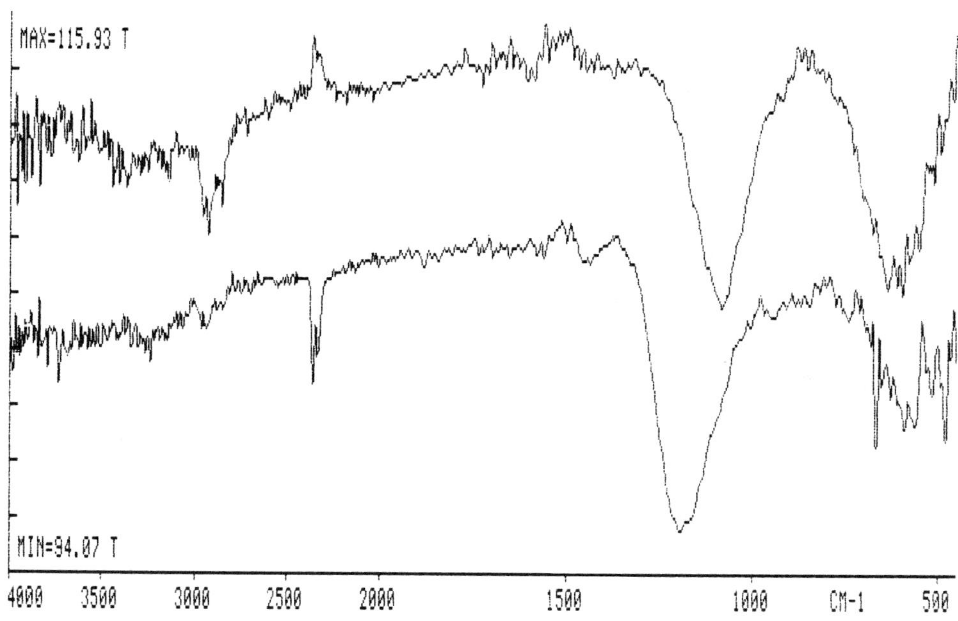

Figure 8. Sample touring - Timken Tour.
Top spectrum: From center of stain.
Bottom spectrum: From center of scar.

easily be seen by a microreflectance tour around the sample, Figure 8. These spectra compared the features in the wear scar with the components on the brown stained area just outside the wear scar. The two spectra show a 116 cm^{-1} shift in the S-O stretching band, differentiating the sulfur interaction with the surface during wear versus the simple metal sulfur contact interaction.

The final spectral comparison to be discussed is the differences between the two wear testers and the deposited material on the surfaces. Figure 9 shows spectra from the center of the wear scar from both a Four-Ball and a Timken sample. Note there were very few similarities between these two spectra. Both samples showed the S-O stretching frequencies, approximately 1200 cm^{-1}. The Four-Ball showed this band to be considerably narrower. The spectrum also showed the evidence of P-O-C and C-H bonds on the surface; whereas, the Timken spectrum only exhibited a small band (946 cm^{-1}) to indicate that the phosphorous additive was reacting with the surface.

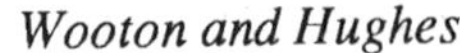

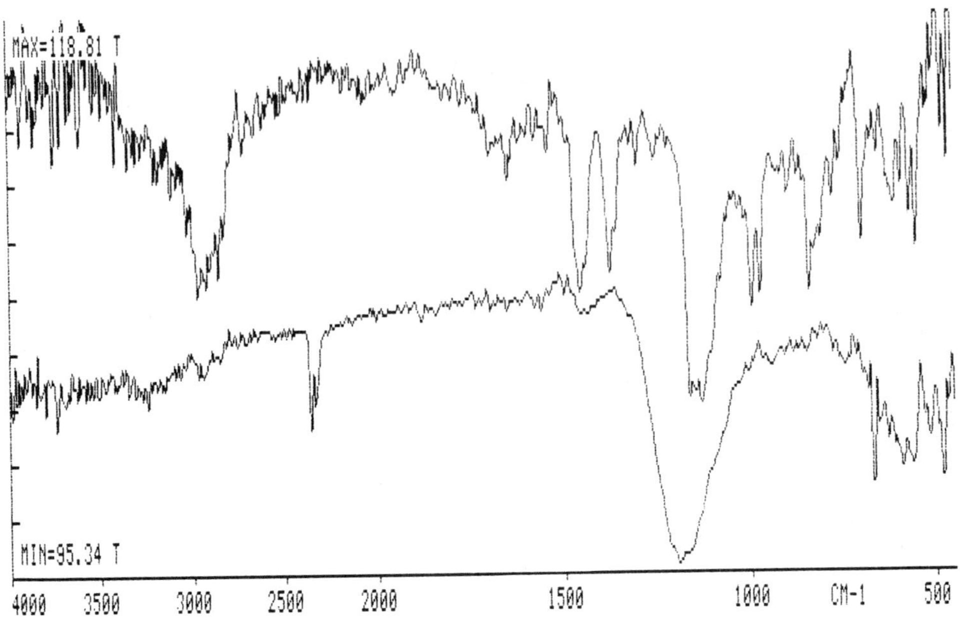

Figure 9. Effects of test configuration - Timken vs Four-Ball.
Top spectrum: Four-Ball specimen - 126 kg (278 lb) loading.
Bottom spectrum: Timken specimen - 50 lb loading.

18.4 CONCLUSIONS

In conclusion, the use of an infrared microscope to show effects of wearing metal surfaces has great potential. The surfaces of both the outside of the wear scar as well as the inside the wear scar can be studied. To do a proper study, it is necessary to carefully prepare and handle the sample prior to and during the analysis. Care must be taken to obtain the correct background to match the sample. Once the proper samples and backgrounds have been developed the remainder of the analyses becomes considerably easier. Since spectral changes could not be seen by varying polarization, detector type or sample surface roughness, these variables do not need to be considered.

With the above techniques, it has been shown that the surface materials can vary inside the wear scar under a number of different test conditions. Two notable conditions that showed spectral changes were sample loading and test configuration. Even with the same lubricant, vast spectral differences could be seen in the different areas in the wear scars. These differences allow one to potentially learn the mechanisms of metal/lubricant interactions during wear, thus leading to the development of lubricants containing improved antiwear properties.

ACKNOWLEDGMENTS

The authors wish to thank Ethyl Corporation for their support and permission to publish this paper. Special thanks are expressed to Spectra-Tech, especially John Coates and Gregg Ressler for their support with some of the equipment used in this study.

REFERENCES

1. *Annual Book of ASTM Standards*, Vol. 5.02, D2782, (1986).
2. *Annual Book of ASTM Standards*, Vol. 5.02, D2783, (1986).
3. D. L. Wooton and D. W. Hughes, *Application of Reflectance Infrared Spectrscopy to the Lubrication Industry*, ASLE, In Press.
4. R. G. Greenler, *J. Chem. Phys.*, 44: 310 (1966).

Index

N

O

S